中国茶经
THE CLASSIC OF CHINESE TEA

于观亭 编著
赖世伟 译

中国出版集团有限公司

世界图书出版公司
西安　北京　上海　广州

图书在版编目（CIP）数据

中国茶经：汉英对照/于观亭编著；赖世伟译.—西安：世界图书出版西安有限公司，2024.3
ISBN 978-7-5232-1078-9

Ⅰ.①中… Ⅱ.①于… ②赖… Ⅲ.①茶文化—中国—汉、英 Ⅳ.①TS971.21

中国国家版本馆CIP数据核字（2024）第052680号

中国茶经（汉英对照）
ZHONGGUO CHAJING（HAN-YING DUIZHAO）

编　　著	于观亭
译　　者	赖世伟
责任编辑	李江彬
出版发行	世界图书出版西安有限公司
地　　址	西安市雁塔区曲江新区汇新路355号
邮　　编	710061
电　　话	029-87214941　029-87236447（市场营销部）
	029-87235105（总编室）
网　　址	http://www.wpcxa.com
邮　　箱	xast@wpcxa.com
经　　销	新华书店
印　　刷	陕西龙山海天艺术印务有限公司
开　　本	787mm×1092mm　1/16
印　　张	31.75
字　　数	550千字
版　　次	2024年3月第1版
印　　次	2024年3月第1次印刷
国际书号	ISBN 978-7-5232-1078-9
定　　价	158.00元

版权所有　翻印必究
（如有印装错误，请与出版社联系）

推荐阅读

推荐人：林楚生
广东省茶文化研究院院长、中国国际茶文化研究会副会长，广东省茶文化研究会会长

《中国茶经（汉英对照）》是在中国茶文化研究空前繁荣、茶产业快速发展的大好形势下出版的第一部中英版新"茶经"。此书在于观亭先生和赖世伟先生的共同努力编著下，吸纳了我国博大精深的茶文化研究成果，内容丰厚，可读性、欣赏性、实用性强。同时，赖世伟先生将其翻译成英文版。这是第一部系统性的普及中国茶叶及茶文化的中英版百科全书，也将是全世界茶文化研究者和爱好者的一本工具书。

中茶院名誉院长、《中华茶通典》执行主编兼《人物典》典长于观亭先生将其一生都奉献给了茶叶。他为实现中国茶业强国梦，使中国茶走向世界，使全世界人喜爱中国茶，了解茶文化，以茶养生，茶和天下做着不懈的努力。他对茶有着纯粹的真诚和炽烈的热情。于老曾说，希望能够看到中国成为真正的全球茶业强国，希望中国茶产业能实现真正的腾飞。

赖世伟先生和于老因茶结缘，并努力实现他们共同的梦想——立志要让中国茶叶重新闪耀世界。赖先生是一位极度热爱茶文化，并能积极投身于茶文化事业中的有为青年。相识至今，他不仅始终高举推广中国茶叶、弘扬茶文化的大旗，更是通过实际行动保持着向世界积极宣传茶文化的使命。赖世伟先生曾跟随我一起走访了各大茶产地，拜访了福建武夷山、福建福鼎、福建政和、湖南安化、云南勐海、云南西双版纳及浙江杭州等地，并向当地的茶叶类非物质文化遗产传承人及当地知名茶企的负责人请教，深入了解不同品类茶叶的制作工艺及品质特性，这些游历，均为其能更加准确地翻译，打下了坚实的基础。这一路，赖世伟

先生始终同我秉持传承创新的理念，为推广中国茶叶，弘扬茶文化竭尽所能。近年来，随着"一带一路"的发展，赖世伟先生始终相信中国茶这一曾经深刻影响并引领世界的饮品，必将会在新时代继续擘画全球的饮品市场。在与赖世伟先生的深入交流中，本人深刻感受到赖世伟先生希望用实际行动去呼应习近平总书记的"茶叙"外交和"中国梦"动员令的热情，事实上他也一直在这条道路上通过实际行动奋勇前行，希望把中国茶和茶文化利用"一带一路"沿线国家和地区的桥梁和纽带作用真正推广出去，让世界真正地了解中国茶文化和中国文化，更好实现中国梦。其在工作期间不仅把中国茶叶推广销往世界各地，也曾主动联系并拜访了多个驻外大使馆，同大使、参赞进行了深入交流，共同讨论中国茶叶及茶文化的传播推广，他们都认同：茶叶的推广需要文化教育做导引。因此，赖世伟先生向这些使馆及外交人员们赠送了《中国茶经》。同时，他还走进孔子学院南方基地，研讨将中国茶叶及茶文化传播到世界各地的多重路径。

　　赖世伟先生深耕于茶文化事业多年，深切感受到现阶段国外有关中国茶叶类文献的缺乏，并且因中西方在茶文化的理解和发展上存在着较大的差异，因而相关文献也需要结合中西方茶文化的差异来进行翻译，才能有利促进茶文化的传承和发展。赖世伟先生具有出国留学经验，其充分利用了自己的语言优势及能力，把通篇十余万字的中文版《中国茶经》翻译成汉英对照版，并大力推动该书的出版。赖世伟先生以弘扬中国茶文化、推动中国茶产业的发展为己任。我希望，这本书可以在让世界人民认识感悟中国茶文化的同时，进而了解中国优秀的传统文化，以此加深我们的文化自信。

推荐阅读

推荐人：王岳飞
浙江大学教授、博导、农学院副院长，浙江省茶叶学会秘书长

  茶、咖啡和可可统称为世界三大无酒精饮料，其中茶的种植和饮用历史最为久远。野生古茶树起源于白垩纪至新生代第三纪的中国西南地区，至今约有6000万至7000万年的历史；国人发现和利用茶叶，可追溯至5000至6000年前的原始母系氏族社会；我国最早关于茶的传说来自《神农百草经》，有"神农尝百草，日遇七十二毒，得荼（茶）而解之"的记载。由此可见，中国茶的历史源远流长。

  在漫长的历史进程中，茶叶成为了中国重要的经济作物，世代茶人不仅积累传承了精湛的产茶技术，还选育出大量的茶树良种。如今，我国的产茶区几乎县县有好茶、乡乡有好茶、村村有好茶，中国因此也是全世界独一无二的名茶品种资源库。随着农业现代化建设与乡村振兴战略的实施，茶叶的栽培与加工工艺得到了科技力量的支持，茶产业也成为了拉动民生的大产业，1085个县、3000多万名茶农靠"一片叶子"生活富足，源远流长的中国茶业被赋予了新时代的重要使命。

  习近平总书记指出："中国是茶的故乡。茶叶深深融入中国人生活，成为传承中华文化的重要载体。作为茶叶生产和消费大国，中国愿同各方一道，推动全球茶产业持续健康发展，深化茶文化交融互鉴，让更多的人知茶、爱茶，共品茶香茶韵，共享美好生活。"伴随着我国改革开放和现代化建设步入新发展阶段，特别是党的十八大以来，中国特色社会主义进入新时代，党中央团结带领全党全军全国各族人民，谱写了经济快速发展、社会长期稳定"两大奇迹"的崭新篇章，中国茶产业进入了新发展时期，实现了一系列突破性进展，取得一系列标志性成果，从茶叶大国走向茶叶强国。"中国茶"成功入选人类非遗以及相关产业数

据表明,中国茶发展呈现良好态势,正在对世界产生巨大影响:2022年,中国茶叶产量318.1万吨,茶园面积4995.4万亩,茶叶消费总量近300万吨,茶叶出口额达20多亿美元。当下中国茶叶产量、茶园面积、消费总量、茶叶出口金额均达到世界第一。真正实现了"中国茶,冠世界"的这一重要成就。

我与于观亭先生结识多年,他还有一个身份是浙江大学茶学系的学生家长之一。于观亭先生曾一手挑起了中国农副土特产品开发公司和中国茶叶产销企业集团两副担子,为中国茶产业发展做出了很大贡献。欣闻青年茶人赖世伟先生与于观亭先生拥有同样的初心情怀,并和他共同完成了第一部汉英对照版《中国茶经》的编写工作,弥补了中国茶叶对外文献的空白。赖世伟先生历尽艰辛,为把《中国茶经》翻译为英文版而不懈钻研突破,期间历经各种曲折与漫长的等待,一直守一颗初心、圆一个梦想。作为青年茶人,赖世伟先生拥有坚定的文化自信与不计回报的奉献之心,矢志不渝地弘扬着中国茶文化,致力于让全世界真正地了解中国茶叶。

中国被誉为茶的祖国和茶文化的发源地,随着我国国际地位的提升,东西方文化不断地碰撞与交融,以茶文化为代表之一的中华传统文化也被赋予了崭新使命。从刚出海时成为上流社会的奢侈饮料,到通过"下午茶"文化在西方社会普及,再到如今与奶类、酒类等融合,文化的碰撞让中国茶名扬海外,可谓"茶和世界,共品共享"。2019年,联合国宣布将每年5月21日确定为"国际茶日";2022年,"中国传统制茶技艺及其相关习俗"被列入人类非物质文化遗产代表作名录——这些重要历史时刻提示了属于茶的时代机遇已经到来,生逢盛世的中国茶人也应尽己之力,深化茶文化的交融互鉴,让更多人知茶、爱茶,共品茶香茶韵,共享美好生活,从而推动全球茶产业持续健康发展。相信此书必定能影响并促进世界对中国茶叶及茶文化的重新理解。

茶起源于中国,盛行于世界。茶的广袤、丰盈、包容,让它自古以来就成为中国与世界沟通联结的方式。当代,以"茶"为桥推动东西方文明交流互鉴,依旧意义非凡。

Introduction

This book is rich in the content of its absorbing results on tea gained by over 30 years of research. It is characterized by readability, appreciation, and practicability. It is truly an encyclopedia that popularizes the tea culture and a reference book for researchers and amateur practitioners.

The Chinese version of The Classic of Chinese Tea was initially edited by Guanting Yu, the Wu Juenong Tea Philosophy Association's vice-chairman, senior engineer, and well-known tea expert. He was born in Shandong province in 1940 and was the engineer of the Tea Husbandry Bureau in the Ministry of Commerce. He has been working on tea development, processing, trade, and industry management for decades, having visited many tea regions and factories in China and the primary tea production and consumption countries. So he familiars with tea production and marketing at home and abroad. Moreover, he combines tea processing technologies with machines and holds a unique idea for tea processing innovation. Therefore, he significantly contributed to the tea processing revolution and the standardization of tea machines.

The English version of The Classic of Chinese Tea was re-edited and introduced by Lai Shiwei (Carl Lai). Carl Lai is deeply in love with tea. He grew up in traditional Chinese and Eastern tea cultures but felt

disappointed and frustrated when he stadied at abroad. Our delicious tea and profound Chinese tea culture have not been popularized as wide as in the world. The prospect of frequent tea consumption providing countless health benefits is far from a new phenomenon. He deems that it is his duty to introduce and promote Chinese tea and culture worldwide. He and other people want to let the world know more about the tea culture and tea customs of China. It seems that the mobility of the Chinese tea culture is slight and almost blank during the past years.

The Classic of Chinese Tea, which is novel in subject matter, fluent in writing style, and comprehensive in content, thoroughly proves that Chinese tea culture has taken on a new look. All this information tells people that this book is characterized by the current times and has a solid cultural fascination. Therefore, it is bound to be a competitive product among the works on tea culture. The publication of this book would positively affect the popularization of the tea culture and the promotion of the tea industry in the world, helping other people to understand the significance of Chinese tea and its background easily.

前言 Preface

茶叶，经过几千年的发展，这个中国传统的经济作物，已由"柴米油盐酱醋茶"的生活必需品发展成为"琴棋书画歌舞茶"的独特文化。如今，茶不但是物质的，也是精神的，对发展农业经济、构建和谐社会起到不可替代的作用。中国是茶及茶文化的发源地，是世界上最早种茶、制茶、饮茶的国家。

Tea is one kind of ancient industrial crop. As one of the seven necessities of Chinese daily lives, including tea, firewood, rice, edible oil, salt, soy sauce, and vinegar, tea was quite common in ancient. Today, it has become a distinctive culture such as lyre-playing, chess, calligraphy, painting, song, dance, and tea after several thousand years. Nowadays, tea is material and carries the human spirit's culture. Therefore, it plays an irreplaceable role in developing agriculture economically and building a harmonious society.

《中国茶经》分为《茶源篇》《茶类篇》《茶造篇》《茶具篇》《茶艺篇》《茶饮篇》《茶风篇》七篇，高度概括了中国茶文化的各个方面和几千年的发展史。

The Classic of Chinese Tea has seven parts: The History of Tea, The Category of Tea, Tea Processing, Tea Ware, Tea Art, Famous Tea, Tea Art, Tea Drinking, and Spread in the World. This book highly generalizes Chinese tea in every aspects and several thousand years of development history.

　　这本中英文版的《中国茶经（汉英对照）》题材新颖、文笔流畅、内容全面，充分反映了中国当前丰富多彩的茶文化新面貌。所有这些，都在告诉人们此书不仅具有时代特色和强烈的文化魅力，也必将成为茶文化著作中的又一精品。此书的出版对弘扬中国茶文化、推动中国茶产业的发展必将起到积极的作用。

　　The Classic of Chinese Tea is innovative, fluent, and comprehensive. It reflects a new character in the colorful Chinese tea culture today. This book has an exceptional cultural charm and will undoubtedly become one more elaborate book of tea culture works. It has a positive impact on Chinese tea culture and this industry.

目录 contents

茶源篇 Tea History / 001

茶经一之源 Chapter 1: The History of Tea / 002

茶的起源 The origin of tea / 004

秦汉茶事 Tea Affairs in the Qin and Han Dynasties / 015

六朝茶事 Tea Affairs in the Six Dynasties / 019

唐朝茶事 Tea Affairs in the Tang Dynasty / 024

宋朝茶事 Tea Affairs in the Song Dynasty / 033

明朝茶事 Tea Affairs in the Ming Dynasty / 043

清朝茶事 Tea Affairs in the Qing Dynasty / 048

茶类篇 The Category of Tea / 057

茶经二之具 Chapter 2: Tea Utensils / 058

茶的分类 Classification of Tea / 062

绿　茶 Green Tea / 073

绿茶的品质 Quality of Green Tea / 075

绿茶的制作 Green Tea Processing Techniques / 082

白　茶 White Tea / 090

白茶的品质 Quality of White Tea / 092

白茶的制作 White Tea Processing Techniques / 097

黄　茶 Yellow Tea / 102

黄茶的品质 Quality of Yellow Tea / 104

黄茶的制作 Yellow Tea Processing Techniques / 111

乌龙茶 Oolong Tea / 115

乌龙茶的品质 Quality of Oolong Tea / 116

乌龙茶的制作 Oolong Tea Processing Techniques / 124

红　茶 Black Tea / 136

红茶的品质 Quality of Black Tea / 138

红茶的制作 Black Tea Processing Techniques / 145

红茶的冲泡 The Brewing of Black Tea / 152

黑　茶 Dark Tea / 157

黑茶的品质 Quality of Dark Tea / 159

黑茶的制作 Dark Tea Processing Techniques / 168

花　茶 Scented Tea / 174

花茶的制作 Processing Techniques / 175

花茶的冲泡 The Brewing Step of Scented Tea / 179

皇　菊 Royal Chrysanthemum / 184

茶造篇 TEA PROCESSING / 187

茶经三之造 Chapter 3: Tea Processing / 188

茶叶产地 Tea Processing Area / 190

江南名茶 Famous Tea in the South of the Yangtze River / 191

江南茶区 The South of the Yangtze River Tea Area / 193

西湖龙井 Xihu Dragon Well Tea (West Lake Longjing Tea) / 197

黄山毛峰 Yellow Mountain Fuzz Tip Tea / 203

洞庭碧螺春 Dong Ting Green Spiral Tea / 206

祁门红茶 Keemun Black Tea / 211

华南名茶 Famous Tea in South China / 215

华南茶区 South China Tea Cultivation Area / 217

安溪铁观音 Anxi Tie Guan Yin Tea / 223

武夷岩茶 Wuyi Rock Tea / 228

武夷大红袍 Wuyi Robe Tea (Da Hong Pao) / 237

白毫银针 White Tip Silver Needle / 240

白牡丹 White Peony White Tea / 244

茉莉花茶 Jasmine Tea / 247

西南名茶 Famous Tea in Southwest China / 251

西南茶区 Southwest China Tea Cultivation Area / 253

滇红功夫茶 Yunnan Province Gongfu Black Tea / 258

云南普洱茶 Yunnan Province Pu Er Tea / 263

老班章生普 Lao Banzhang Unfermented Pu Er Tea / 266

江北茶区 The North of the Yangtze River Tea Area / 276

茶具篇 Tea Ware / 281

茶经四之器 Chapter 4: Tea Ware / 282

茶　器 Tea Utensil / 284

备水器 Water Ware / 286

理茶器 Preparing Ware / 290

置茶器 Tea Setting Tools / 294

品茗器 Tea Savoring Utensils / 300

洁净器 Utensil Cleaning Tools / 306

茶具的种类 Types of Tea Ware / 311

气韵独特的紫砂茶具 Unique Purple Sand Clay Tea Wares / 312

温润细腻的瓷器茶具 Smooth and Delicate Porcelain Tea Wares / 321

淳朴典雅的漆器茶具 Elegant and Straight forward Lacquer Teawares / 332

华贵不实的金玉茶具 Gimcrack Precious Stone and Metals Tea Wares / 335

通透夺目的玻璃茶具 Translucent and Attractive Glass Tea Wares / 342

自然粗犷的竹木茶具 Natural and rugged Bamboo Tea Wares / 345

茶具的选用 Tea Wares Selection / 347

茶艺篇 Tea Arts / 353

茶经五之煮 Chapter 5: Tea Brewing / 354

水 Water / 356

讲究的泡茶水 Fastidious Selection / 358

天下名泉 Famous Springs in China / 363

境 Ideal State / 365

品茗佳境 Excellent Environment for Drinking Tea / 367

意境之美 The Beauty of Artistic Conception / 371

人 People / 379

  艺 The Arts / 381
    茶艺的精神内涵 Spiritual Connotation of Tea Art / 383
    茶艺的种类 Type of Tea Arts / 386
    分类的方法 Classification Method / 388
    茶的鉴别 Identification of Tea / 394
    茶的鉴赏 Appreciating Tea / 396

茶饮篇 Tea Drinking / 405

茶经六之饮 Chapter 6: Tea Drinking / 406
  茶与健康 Tea and Health / 408
  茶的健康元素 Tea Health Element / 409
  茶的保健功效 Health Effect of Tea / 418
  科学合理地饮茶 Drinking Tea Scientifically / 426
  茶叶的保存 Preservation of Tea / 428
  影响茶叶品质的因素 Main Factors Affecting the Quality of Tea / 430

茶风篇 Spread in the World / 437

茶经七之事 Chapter 7: Spread in the World / 438
  茶之事 Tea Story / 440

茶经八之出 Chapter 8: Spread in the World / 444
  在世界上传播 Spread in the World / 446

茶马古道 Ancient Tea Route / 448

传入日本 Introduced into Japan / 452

来到欧洲 Spread to Europe / 456

茶在英国 Tea in Britain / 462

异域茶情 Tea Culture of Foreign Lands / 468

朝鲜、韩国茶礼 Tea Etiquette of North Korea and South Korea / 470

日本茶道 Japanese Tea Ceremony / 472

土耳其茶事 Turkish Tea Story / 477

英式下午茶 British Afternoon Tea / 479

摩洛哥茶饮 Morocco Tea Drinking / 484

茶经九之略 Chapter 9: Tea Tool Omitted / 488

茶经十之图 Chapter 10: Tea Flipchart / 490

茶源篇
Tea History

天赐香叶，带着日之热烈、月之温润、风之灵动、雨之醇香，是以为茶。人间有神农氏，采其为饮，于是人间便多了一种沁人心脾的仙露。

The beautiful tea leaf is a godsend. It absorbs sunlight from the Sun, warmth, humidity from the Moon, and living energy from the Wind, the pure and aromatic smell from the Rain. Since Shen Nong (Divine Husbandman) picked up tea leaves and drank them, there was one kind of drink to gladden the heart and refresh the mind.

The tea tree is a precious tree species in southern China. The height of the trees varies from one or two feet to tens of feet. Trees grow around the Bashan and Xiachuan areas. (Currently Eastern Sichuan Province and Southwest Hubei Province). Some trees are so large that two adults can hold them together. You must cut branches off the trees to pick up the bud leaves. The tea tree's shape looks like a bottle gourd, the tea leaf looks like a gardenia, the flower looks like a cottage rose, the seed looks like a palm, the stalk looks like a clove, and the root is like a walnut. There are three kinds of writing with characters from ancient times. And its five names were: Cha, Jia, She, Ming, and Chuan, of all ages.

种茶的土壤，以岩石充分风化的土壤为最好，含有碎石子的砾壤次之，黄色黏土最差。一般说来，如果茶苗移栽的技术掌握不当，移栽后的茶树很少长得茂盛。茶树苗一般生长三年即可采茶。茶叶的品质，以山野自然生长的为好，在园圃栽种的较次。在向阳山坡，林荫覆盖下生长的茶树，芽叶呈紫色的为佳，绿色的略差；芽叶以节间长，外形细长如笋的为好，芽叶细弱的较次。叶绿反卷的为好，叶面平展的次之。生长在背阴的山坡或山谷则品质不好，不值得采摘。因为它的性质凝滞，喝了会使人胀腹。

Rock weathering soil is the best culture substrate for tea. The ground containing rock fragments comes second. The worst is the yellow clay. In general, tea trees will not flourish without mastered transplanting technology. You can pluck and then drink tea in three years as usual. The wild tea is the best, then the garden planting tea. Tea leaves grow in the shade of the tree on the hillside, and the purple bud tea is good. The purple bud tea variety is better than the green leaf variety with the long and thin internode length. They look like bamboo shoots, better than slim and thin bud tea. Those good quality tea leaves are often revolute and shrinking. The thin and flat leaves are next to it. Tea, which grows on the shady hill, is worse. It is not worth being picked. It may causes stomach pains.

茶的起源
The origin of tea

　　中国是茶的故乡，也是世界上最早种植和利用茶的国家，茶叶伴随着古老的中华民族走过了漫长的岁月。打开中国五千年的文明发展史，几乎从每一页中都可以嗅到茶的清香。茶不仅是一种饮品，更是一种博大精深的文化。茶文化是中国传统文化的重要组成部分，是中华文明长河中的一颗璀璨明珠。

　　唐代陆羽的《茶经》不仅系统地总结了种茶、制茶和饮茶的经验，而且将儒、佛、道三教思想与中国古典美学的精髓融入茶事中，把茶事活动升华为一种富于中华民族特色的高雅文化，即中国茶文化。

　　China is the hometown of tea and the first country to plant and use it tea. The ancient Chinese nation accompanies a long history of traditional Chinese tea culture. You can enjoy the fresh smell of tea in every period of five thousand years of Chinese civilization history. Tea is not only a drink but a broad and profound Chinese culture. Chinese tea culture is an essential part of Chinese traditional culture and a bright pearl of the long history of Chinese civilization.

Lu Yu wrote The Classic of Tea of the Tang Dynasty (618 A.D.–907 A.D.). It systematically summarized the experience of planting, processing, and drinking tea and integrated the traditional views on Confucianism, Buddhism, and Taoism and the essence of Chinese classical aesthetics. As a result, tea activity develops distinctive national traits, high culture, and popular cultural traditions.

美丽的传说
A Beautiful Legend

陆羽《茶经》里说："茶之为饮，发乎神农氏。" 传说中的炎帝神农氏是茶的发现者，同时他也是传说中发明药物来治疗疾病的人。

In his book The Classic of Tea, Lu Yu said tea started to be a drink in Shennong. According to the tale, Shennong (the Yan Emperor) was the tea discoverer who invented medicine to cure diseases.

神农氏为了辨别草物的药理作用，曾经亲口品尝百草。有一次他在野外考察休息时，用釜锅煮水，恰巧有几片叶子飘落进来，锅里的水变成黄绿色。神农氏不以为然，喝了一点其中的汤水，却惊奇地发现，这黄绿色的水味道清香，竟是一味不可多得的饮品。随着时间的推移，神农氏得出了这种植物具有解渴生津、提神醒脑和利尿解毒的作用。至于"茶"的名字的来源，也和神农氏有关。传说中的神农氏长着一个玻璃一样透明的肚子，但凡是吃进肚子里的食物都能够被看得清清楚楚，因此能够知道这种食物对于身体的利弊，这也是他多次中毒不死的原因。他喝了黄绿色的水之后，看见这种水在肚子里流淌，所到之处，肠胃被"擦洗"得干干净净。于是他就把这种植物叫做"擦"，后来就转化为"茶"的发音。

Shennong tasted hundreds of wild herbs to compare their pharmacological action to find remedies to treat illnesses. He was doing a field trip and boiling water in the kettle. Just then, a few leaves were falling into the kettle. Then the water turned yellow-green. He did not care much about it. But when he drank this yellow-green water, he was surprised that the water had a fresh taste. It was a rare medicinal material. As time went by, Shennong concluded that this plant has some effects on quenching thirst, producing saliva, refreshing one's mind, and promoting diuresis to detox the body. So Shennong created the name for tea. According to the tale, Shennong had a clear glass-like belly. Everything in his transparent belly would be visible. So he could see the pros and cons of this kind of food in the body. That's why he could survive after being poisoned. The water was rushing in his stomach. Everywhere the water went, the stomach was brushed clean. The word 'brush' in Chinese has a similar pronunciation to "cha." So he called this plant "cha." So then, the pronunciation of tea is "cha."

关于产地的争论
The Argument about the Country of Origin

茶树原产于中国,这是被举世公认的,但是在19世纪初,一位英国少校在印度发现了野生的大茶树,于是有人开始认为茶的发源地是印度而非中国,从而在国际学术界引发了一场争论。

It is a universally recognized fact that tea is native to China. But a British major found the wild tea camellias in India in the early 19th century. As a result, there are some rumors around. India is the home of tea instead of China, thus sparking a debate in the international academic community.

1823年,英军少校布劳士(R.Brouce)在印度与缅甸的交界处发现了一株高约13米,直径约1米的野生古茶树。次年,他的哥哥在印度境内也发现了类似的野生茶树,于是他们据此断言,印度是茶的原产地。之后,很多西方学者都坚持这一观点。

In 1823, a British Major General R.Brouce found a wild tea tree about thirteen meters high and one meter in diameter. His brother also found some similar wild tea trees in India. So they affirmed that India is the origin country of tea. Afterward, lots of Western scholars hold this view.

1919年,荷兰学者斯图尔特认为,茶叶的原产地分为两种:大叶种原产自印度、缅甸和中国云南;小叶种则产自中国东南部。1935年,美国学者威廉·乌克斯在其著作《茶叶全书》中又提出了茶叶原产地的"多元说",认为茶叶原产自印度和中国,以及泰国、缅甸等国家和地区。除此之外,仍有很多国家的学者坚持着茶叶发源于中国的观点。

In 1919, a Dutch scholar, Stewart, considered two countries of origin of tea. The large-leaf tea is native to India, Myanmar, and the Chinese Yunnan Province. The small-leaf variety of tea is native to southeast China. In 1935, an American academic put forward the theory about tea's various origins in All about Tea. The tea plant originates from India, China, Thailand, Myanmar, and other countries and regions. But beyond that, many scholars from countries insist on the view that tea originated in China.

最初的记载
The Historical Record

在周武王伐商灭纣时,参加征战的巴蜀等南方小国部落就把茶作为贡品敬献给周武王。晋常璩著的《华阳国志》中记载:"周武王伐纣,实得巴蜀之师,……茶蜜……皆纳贡之。"武王伐纣的时间约在公元前1066年,由此可见,中国有明确记录的茶事活动距今至少已有3000年的历史了。

According to the records, many tribes and small states in the Bashu area who participated in the war once brought tea as a tribute to King Wu of Zhou when he crusaded against the State of Zhou and Shang Dynasties. So King Wu got those Bashu troops and tea and honey, recorded in Chronicles of Huayang. King Wu successfully overthrew the Shang Dynasty in the year 1066. Thus, some individual records of Chinese tea affairs have been documented for at least 3000 years.

现在所能够看见的文献资料里面，有着确切的茶的记载的，最早并且最可靠的应该是汉代王褒所撰写的《僮约》。这篇文章写作的时间是汉宣帝神爵三年（前59），是茶学史上重要的文献。其中的"烹茶尽具""武阳买茶"，说明"茶"已经成为当时社会饮食的一项，并且是用来待客的贵重之物，饮茶已开始在中产阶层中流行。

Tong Yue is the earliest and most reliable recording of tea in the existing document literature. Wang Bao wrote this book during the Han Dynasty. This article is the critical literature in the long tea history, 59 B.C. It detailed that tea wares should be well prepared and be washed clean. People went to Wuyang County to buy tea for home drinking frequently. Wuyang County was located in Shuangjiang Town, Pengshan County, Chengdu. At that time, tea has become a part of people's diets. People began to serve valuable tea to their friends. Drink tea is gaining in popularity among middle-class Chinese.

文物的明证
Related Cultural Relics

中国拥有世界上最古老的与茶相关的古代文物，从另一个角度为中国是茶树起源地的观点提供了明证。

China has a lot of precious and oldest ancient relics related to tea in the world. Therefore, it provides evidence from a different perspective that China originates tea.

近年来，在浙江省上虞市出土的东汉时期的瓷器中，有壶、盏、杯、碗等器具。据考古学家判断，这些器物当属世界上最早的茶具。这说明东汉时期饮茶已渐渐普及。湖北省江陵县的西汉古墓中还曾出土过一些作为陪葬品的茶叶；湖南省的长沙马王堆汉墓中也曾出土过一只刻有"茶"字的青瓷瓮，这被考古学家推定为是当时人们用来贮存茶叶的器具。

Some utensils such as pots, Zhan (tiny teacups), cups, and bowls emerged during the Eastern Han Dynasty. They were unearthed in Shangyu city, Zhejiang province. Archaeologists determined that these were the world's earliest tea sets. Drinking tea had become very common in the Eastern Han Dynasty. As part of funerary objects, tea was also found in the West Han Dynasty, Jiangling County, Hubei Province.

A celadon urn engraved with 'tea' was also found in Mawangdui Han Tombs of Changsha City, Hunan Province. Some archaeologists presumed that there might be some utensils for storing tea. The Chinese characters 'si' was found in archaeology's detailed list of funerary objects. The word 'Si' and 'Jia' are the same. Therefore, it was checked and testified by some archaeologists.

中国野生茶树的发现
The Discovery of Wild Tea Camellias in China

在中国古代的著作中曾经有很多关于野生茶树的记载，如公元6世纪以前的《桐君录》中提到的"瓜芦木"即为茶树的大叶变种；唐代陆羽的《茶经》中明确记载了"茶者，南方之嘉木也"；宋代沈括的《梦溪笔谈》中也有"建茶皆乔木"；明代《大理府志》记载"点苍山……产茶树高一丈"，等等。

There are many records about wild tea camellias in some authoritative professional books. For example, the Gua lu Mu was a large-leaf variety of tea camellias in the book of *Tong Jun Lu*. Tea tree is a premium timber in southern China. *The Classic of Tea*, edited by Lu Yu from the Tang Dynasty, documented this information. Shen Kuo wrote the book of *Meng Xi Bi Tan* and also mentioned that the type of tea camellias were all trees in Jianxi County in the North Song Dynasty. In the Ming Dynasty, *Da Li Fu Zhi* recorded that there were a lot of ten feet tea trees in Diancang Mountain.

除了史书的记载，研究人员于1939年至1940年在中国贵州务川先后发现了十几株家生大茶树；1958年，在云南发现了高约10米，树龄已有800多年的"茶树王"；1961年更是发现高达30多米，树龄约1700年的家生茶树；同一时期，在广东、广西、四川、湖南等10个省区的198处发现家生大茶树，大茶树是如此之多，分布如此广泛，堪称世界之最。

Researchers found more than ten domestic tea plants in Wuchuan City, China, Guizhou Province. The king of tea trees and 800-year-old was about ten meters high in Yunnan Province in 1958. Some 1700-year-old domestic tea trees, about thirty meters high, also were discovered in 1961. In the same era, domestic tea plants were found in 198 places in more than ten provinces: Guangdong, Guangxi, Sichuan, Hunan, etc. In addition, tea plants were the most widely distributed in many areas of China globally.

中国是茶树的原产地
Native to China

当然，发现野生茶树的地方，不一定就是茶树的原产地。中国是茶树的原产地的结论是科学家们依据现实，从各个方面分析考证后而得出的，在当今世界已再无争议。

根据植物学家和地质学家的分析，茶树起源至今已有6000万年至7000万年的历史了。印度所处的喜马拉雅山南坡在那个时期被埋在海底，不可能生长茶树，而在中国西南地区发现的山茶属植物有100余种，可以推测这里是这一植物区系的起源中心。

Of course, it does not necessarily mean that the wild tea has been found in the origin area of tea. The conclusion is now beyond doubt that from the scientists. China is the origin of the tea tree through the study in all of its aspects. According to analysis, tea has 60 million to 70 million years of history from botanists and geologists. Because India was in the deep seabed south of the Himalayas, it was impossible to grow tea. In China, people discovered more than 100 kinds of Camellia in the southwestern part of China. We can determine that the southwestern China area is the original center of this plant flora.

茶树起源于中国，距今已有 6000 万～7000 万年。茶被人类发现和利用，约有四五千年的历史。最初人们将茶树叶放在水中煮饮，茶汤被当作药用，食嫩叶作为蔬菜。随着时间的推移，茶慢慢普及成为一种饮品。

Tea trees originated in China. It has been more than sixty to seventy million years. And it was found and has already been used for four to five thousand years. At first, people brewed tea leaves in water. Tea soup was for medicinal purposes, and those tender leaves were eaten as a vegetable. As time goes by, it is gradually popularized tea as a kind of drink.

在中国，茶文化的发展历程大致经过了"发乎于神农，闻于鲁周公，兴于唐而盛于宋"的过程。茶文化经历了秦汉的启蒙，魏晋南北朝的萌芽，唐代的确立，宋代的兴盛和明清的普及等几个阶段。

In China, tea's cultural development has generally passed the following historical phases: the discovery phase from Shennong, and it has been popular since the Lu Zhou Gong in 1100 B.C. It became popular in the Tang Dynasty and flourished in the Song Dynasty. The tea culture of China experienced the enlightenment phase of the Qin and Han Dynasties. However, it was still embryonic during Wei Jin Southern and Northern

Dynasties, establishing the Tang Dynasty, the Song Dynasty's prosperity, and the popularization in the Ming and Qing Dynasties.

茶文化的发展历程不仅仅是一种饮食文化的形成过程，同时也折射出中华民族上下五千年积淀下来的精神特质与文化内涵。那么，就让我们来亲身体验一下茶叶历史的变迁吧！

The development history of Chinese tea culture is to establish food culture and reflect the spiritual and cultural connotations. It accumulated a rich culture over five thousand years of Chinese history. So, let us realize the endless charm of the long history of tea.

秦汉茶事
Tea Affairs in the Qin and Han Dynasties

茶源篇 Tea History

巴蜀茶风
Tea Story in Bashu Area

　　巴蜀自古被人们称为孕育中国茶业与茶文化的摇篮，古代的巴蜀国也可以说是中国最早的产茶地区。明代杨慎的《郡国外夷考》中记载："《汉志》葭萌，蜀郡名，萌音芒。《方言》，蜀人谓茶曰葭萌，盖以茶氏郡也……"表明很早以前，蜀人已用"茶"来为当地的部落和地域命名了；同时也反映出巴蜀地区在战国之前已经形成了具有一定规模的茶区。明末学者顾炎武在他的《日知录》中提到"自秦人取蜀而后，始有茗饮之事"，也反映了茶饮是秦国统一巴蜀之后开始传播的。

　　Bashu area is the cradle of the ancient Chinese tea industry and culture. Ancient Bashu was the earliest tea-producing area in China.

Jun Guo Wai Yi Kao, written by Yang Shen in the Ming Dynasty, has documented that Jia Meng was the original name of Bashu County. In the Bashu area, people called tea "Jia Meng," and the tea county was also named. It indicated that the Bashu people had already used tea to document the local tribes and regions. At the same time, it reflects a certain amount of tea cultivation plantations in the Bashu area before the warring states. Gu Yanwu wrote Ri Zhi Lu. It mentioned that some drinking tea activities also began to spread after Qin unified China in 221 B.C.

西汉时，王褒的《僮约》中已有"烹荼尽具"和"武阳买荼"的记载。可见，在当时的巴蜀地区，饮荼已广泛盛行，茶叶甚至成为了一种商品。三国时期，魏国《广雅》一书曾记载："荆巴间采茶作饼，成以米膏出之……"反映出巴蜀地区独有的制茶方式和饮茶方法。

Tong Yue, written by Wang Bao in the Western Han Dynasty, is the oldest book with rich records about tea sets. It should be prepared and washed clean. Then, people should go to Wuyang County to buy tea for home drinking. Thus, drinking tea had already been very extensive at that time. And tea has even become a commodity. The book Guang Ya in the state of Wei in the Three Kingdoms recorded that people picked up the tea leaves to make tea cakes, then put thick rice syrup on them. It shows a unique method of making and drinking tea in the Bashu area.

煮饮法
Brewing Method

所谓煮茶法，是指将茶放在水中烹煮而饮。唐代以前没有制茶法，从魏晋南北朝一直到初唐，人们主要是将茶树的叶子采摘下来直接煮成羹汤来饮用，饮茶就像今天喝蔬菜汤，吴国人称此为"茗粥"。

By "Brewing method", we mean putting tea leaves into the water and then boiling them for drinking. Before the Tang Dynasty, there were no processing methods to process tea. From Wei Jin Southern and Northern Dynasties to the early Tang Dynasty, people mostly picked off the leaves then boiled them to make a soup for eating directly. Drinking tea was just the same thing as having vegetable soup today. So the people of Wu Kingdom called it tea congee.

唐代中后期的饮茶方式以陆羽式煎茶为主，但煮茶的习惯并没有完全被摒弃，特别是在少数民族地区较为流行。晚唐樊绰的《蛮书》曾记载："茶出银生城界诸山，散收，无采早法。蒙舍蛮以椒、姜、桂和烹而饮之。"这表明唐代时期，煮茶往往要加入盐、姜等各种佐料。

It was mainly Lu Yu style Sencha popular in the Mid-and-late Tang Dynasty. But people have not forsaken the habit of making tea. . It was widespread, especially among some northern minorities. For example, fan Chuo's Man Shu recorded tea from Jie Zhu Mountains in Yin Sheng Cheng (currently Jing Dong County in Yunnan Province). The making method was

easy. I just dried the tea leaves on the ground. People in Mengshe, located in the Wei Shan and Nan Jian area of Yunnan Province, boiled and drank tea with prickly ash, ginger, and cinnamon. It shows that people often cook tea with salt, ginger, and some seasoning.

到了宋朝时期，北方少数民族地区在茶中放入盐、干酪和姜等一起烹煮，南方地区也仍然保留有煮茶的习俗。明清至今，煮茶法主要是在少数民族中流传使用。

Until the Song Dynasty, China's northern minorities boiled tea with salt, cheese, and ginger. And beyond that, there were some customs for cooking tea in Southern China. During the Ming and Qing Dynasties, ethnic minorities still used the cooking tea method.

六朝茶事
Tea Affairs in the Six Dynasties

茶源篇 Tea History

重心东移
Spreading Eastward

三国两晋时期，长江中下游地区因为便利的地理条件和较高的经济文化水平，茶业和茶文化也得到了较大发展。该地区在中国茶文化传播中的地位，逐渐明显且重要起来，呈现出取代巴蜀之势。此外，由于六朝基本上都是定都建康，中国茶业的重心也逐渐由西向东移，从而使得中国南方，特别是江东的茶文化和饮茶习俗有了较快的发展。

The tea industry and tea culture significantly developed in 222 – 589. It mainly resulted from a convenient geographical position and a higher economic and cultural level during the Three Kingdoms period and the Jin Dynasty. Therefore, this area became more significant for promoting Chinese tea culture and showed a clear trend to replace the Bashu area's position.

Furthermore, as the Six Dynasties' emperor's capital was in Jiankang City, the Chinese tea industry's center gradually moved from west to east. So the tea culture and tea-drinking customs developed fast and well in southern China, especially south of the Yangtze River.

从药用到饮用
Change from a Chinese Herbal Medicine to a Drink

秦汉时期，茶并非普通百姓的日常饮品，而是更多地以其药用价值出现在人们的生活中。

据史料记载，到了三国时期，茶开始在王室贵族等上层社会流行。两晋和南北朝时期，茶被作为药用还是日常饮用，因南北地域和习俗的不同，而经历了一段具有南北差异的过渡期。

Tea was not the daily drinking of ordinary people in the Qin and Han Dynasty but had medicinal effects. According to the historical record, tea became a fashion among some upper-class and noblemen. It experienced a transition period with an apparent geographic difference and social institutions difference in the south-north region of the Western Jin and Eastern Jin Dynasties and the Southern and Northern Dynasties.

由于茶叶原产自云南、四川等地，南方饮茶习俗较北方成熟略早。南下的中原贵族逐渐适应了南方的饮茶文化并喜欢上了饮茶。而东晋南渡之初，北伐志士刘琨在信中写道："前得安州干姜一斤，桂一斤，黄芩一斤，皆所须也。吾体中溃闷，常仰真茶，汝可置之。"可见，北方士族依然将茶视为药饮。

Tea is native to Yunnan, Sichuan, and other places. So the more mature tea-drinking customs experience was in the south than in the north. The Central Plains nobility moved to the south. They gradually got used to the tea culture and then loved drinking tea. While in the beginning, the imperial clan of the East Jin Dynasty moved to the south. The patriot in the Northern Expedition, Liu Kun, wrote in a letter. He got one catty of ginger, one catty of cinnamon, and one catty of Baikal skullcap in Anzhou County. The feeling of fatigue and depression could be relieved with tea. It is thus clear that the Northern Intelligentsia also regarded tea as medicine.

客来敬茶
Serving Tea

中国自古以来就有以茶待客的传统习惯。而以茶待客的风气，最早可以推溯到两晋南北朝时期。东晋时期，茶饮已经成为三吴（吴郡、吴兴、会稽）地区和建康（今江苏南京）一带常见的待客之物。

Since ancient times, China has had a good tradition of serving tea to its friends. This custom can be traced back to the age of the Wei Jin Southern and Northern Dynasties. Tea drink was a typical treat during the Eastern Jin Dynasty.

茶文化的萌芽
The Birth of Tea Culture

至魏晋时期，饮茶的方式逐渐进入烹煮的阶段，对烹煮的方法技巧也开始讲究起来。饮茶的形态除了在种类上呈现多样化的特点之外，还开始具有一定的仪式、礼数和规矩，人们日益自发自觉地规范遵守起来。

Tea drinking gradually entered a new development phase in the Wei and Jin Dynasty periods. The tea-drinking becomes increasingly diversified in identity and has a specific ceremony, etiquette, and custom. People follow the rule day by day consciously. The etiquette was both formal and standard.

在这一时期，茶也开始成为文人雅士吟咏、赞颂和抒情达意的对象。杜毓的《荈赋》、左思的《娇女诗》等作品从各个方面对种茶、煮茶、饮茶等茶事都进行了描述。

During this period, tea became an appreciation and glorification object and expressed the thoughts and feelings of spiritual pursuit from scholars and gentlemen. Du Yu's article "Fu" and Zuo Si's poetry "Jiao Nv Poem" depicted tea activities in detail, such as planting tea, cooking tea and drinking tea, and so on from every aspect.

此外，茶作为一种健康的饮品，其清香雅致的特质被赋予高雅淳朴的精神力量，与儒、佛、道和神、鬼、怪等联系起来，开始进入宗教领域。

Furthermore, as one kind of healthy drinking, its delicate fragrance and elegant quality were endowed with elegance, simplicity's spiritual strength to relate the Confucians, Buddhas, Taoism, God, Ghosts, and Monsters. Then tea started entering the fields of religion.

从茶文化发展史的整体看来，虽然这一时期的中国茶文化还仅仅处于发展的萌芽阶段，茶风还没有普及到民间百姓，人们饮茶更多地关注于茶的物质属性和药性，而不是其文化功能，但是仍为后世茶文化的发展和完善奠定了一定的基础。

Although Chinese tea culture during that period was still in the stage of embryonic development and tea stories had not been popularized in the folks without more attention to the physical attributes and harmonic numerous medicine properties of medicine rather than its cultural function, the tea culture of that period had laid a solid foundation for the development and improvement for posterity.

唐朝茶事
Tea Affairs in the Tang Dynasty

比屋之饮
Drink Tea from Door to Door

"比屋之饮"是指唐朝时期饮茶已经十分普遍，特别是在唐都长安，茶饮习俗几乎走进每家每户的意思。唐朝时期的经济发展日趋繁盛，文化昌明，社会处处生机，充满活力。这些有利条件为包括茶业在内的各行各业的发展提供了动力。茶圣陆羽就生于这样一个繁荣的时代。

The allusion to drinking tea from door to door indicates that drinking tea was familiar to most families in Chang An(now Xi'an) in the Tang Dynasty. Economic and cultural development had become increasingly prosperous and flourished during the Tang Dynasty. As a result, society was full of vitality and energy. These favorable conditions provided the motive power for all trades, including the tea industry. The saint of tea Lu Yu was born in this prosperous dynasty.

特别是唐朝中期以后,饮茶之风已经开始从皇宫、贵族、文人雅士阶层逐渐普及到社会中下阶层,特别是得到了普通百姓的欢迎。唐代开元年间(713-741),社会上茶道兴盛,饮茶之风大兴。史料记载,文成公主入藏时(641)就曾把茶叶及茶籽随身带入吐蕃,饮茶使得以肉食为主的藏民获益良多。很快,饮茶习俗在西藏地区逐渐形成,发展到今日"宁可三日无粮,不可一日无茶"的程度。

Tea-drinking customs gradually spread from upper-class statuses, such as royal families, aristocrats, and refined scholars, to the lower-middle class, especially among ordinary people after the Tang Dynasty. During the Kaiyuan perid of Tang Dynasty. Tea ceremonies and drinking tea customs were famous in society. Historical records show that Princess Wencheng had already brought tea and tea seeds when she married Tibet in 641. The Tibetans, who have a carnivorous diet, benefited from drinking tea. The drinking tea custom gradually formed in Tibet very soon and had progressed to a certain degree that people could not live without food for three days rather than a day without tea.

文成公主与茶
Princess Wen Cheng and Tea

唐朝时文成公主远嫁吐蕃,促进了汉藏两个民族之间的友好和经济文化交流。由于文成公主酷爱饮茶,嫁妆里自然也少不了茶叶,茶文化也随之传入西藏,并在当时的贵族间盛行,因此开始了两地之间的茶马交易。

During the Tang Dynasty, Princess Wen Cheng married Songzan Ganbu in the Tibetan regime. This marriage promoted friendly relations and economic and cultural exchanges between Han and Zang nationalities. Princess Wen Cheng loved tea. Therefore, tea must be in her dowry, tea culture, and Princess Wen Cheng to Tibet. It flourished among the nobility. Thus, it started the Tea-horse Ancient Trading between the two places.

相传,藏区人民最爱喝的酥油茶也是由文成公主创造的。当时,文成公主刚嫁到吐蕃,适应不了高原干冷的气候环境,对每餐肉多、奶多的饮食方式不习惯,常常感到油腻,不好消化。于是便想到把清爽的茶加进奶中饮用,果然好了很多,这便是奶茶的由来。她还尝试在煮茶时,加入酥油、盐、松子等,发展成了现在的酥油茶。文成公主还经常把茶赐予臣民,使得越来越多的藏民感受到茶水清幽的口感和醒脑提神的功效,对西藏茶叶的传播和发展作出了巨大的贡献。

According to legend, the buttered tea was adored by Tibetan people and created by Princess Wen Cheng. She could not adapt to the plateau's cool-dry climate and was not accustomed to the Tibetan diet, with much more meat and milk for every meal. She always felt too oily, and her digestion was worse than before. Thus she came up with the idea: mix the crisp flavors of tea with milk. Sure enough, it turned out better. That was the origin of tea with milk. She also tried adding butter, salt, and pine nut into tea, then found it today's buttered tea. Princess Wen Cheng often gave tea to her people as a gift. It also made more and more Tibetans addicted to the pure and elegant taste and experienced refreshing the mind. She made significant contributions to promoting and developing tea culture in Tibet.

茶 制
Laws and Regulations of Tea Economic

唐朝时期，茶叶生产得到了较大发展，从事茶叶买卖的商人大多迅速致富。但唐中期以后国家却出现了财政危机，在这种形势下，唐王朝开始制定关于茶叶的经济法规，以增加财政收入，这些法规包括税茶、贡茶、榷茶、茶马互市等，大多被历代沿袭下去并成为定制。

税茶：唐德宗建中元年（780），户部侍郎赵赞提议朝廷对茶征收10%的税。贞元九年（793），张滂据此创立了税茶法。

Tea production had significantly developed during the Tang Dynasty. All traders who engaged in tea sales could get rich quickly. But the goverment had a financial crisis after the middle Tang Dynasty. People began to make economic rules and laws on tea to increase the Tang Dynasty revenues in a situation like that. Some regulations included tea tax collection, tribute tea, levy tea taxes, Tea-Horse trade centers, etc. Most of those regulations were inherited and continued to become law. Tea tax collection: The assistant minister of the Ministry of Revenue, Zhao Zan, proposed a 10% tax on tea. Then Zhang Pang initiated the establishment of the Tax Tea Law.

榷茶：榷的本义为独木桥，引申为专卖或垄断。唐武宗时期，茶叶开始"禁民私卖"，榷茶制度正式确立。

Levy indirect taxes on tea (Que Tea): Single-plank Bridge was the original meaning. Levying indirect taxes is the monopoly. The government started to ban the illicit sale of tea during Emperor Wuzong of Tang. It symbolized the formal establishment of the levying indirect taxes system.

贡茶：贡茶不是商品，而是专供朝廷使用的茶叶。由于其制作精致讲究，大大推进了种茶和制茶技术的进步。但同时贡茶也加重了茶农的负担，并在一定程度上阻碍了茶叶贸易的发展。

Tribute tea: Officials must pay tribute tea to the royal. Their delicate and elegant processing broadly promoted technology's advancement in planting and manufacturing tea. But it also increased the burden on farmers and somewhat delayed the tea trade.

煎茶法
Sencha Tea Method

煎茶法主要是指陆羽在其编写的《茶经》中所记载的一种饮茶方法。煎茶法通常应用于饼茶，主要程序有备茶、备水、生火煮水、调盐、投茶、育华、分茶、饮茶、洁器共九个步骤。煎茶法一出现就受到士大夫阶层、文人雅士和品茗爱好者的喜爱，特别是到了唐朝中后期，逐渐成熟并且流行起来。

由于茶圣陆羽是煎茶法的创始人，因此煎茶法又被称为"陆氏煎茶法"。煎茶之道可以说是中国茶道的雏形，兴盛于唐朝、五代和两宋时期，历时约500年。

The Sencha Tea method is mainly referred to as a drinking way recorded in Tea's Classics, edited by Lu Yu. People always used brick tea as the material for this method. The whole process includes nine steps: prepare tea, prepare water, boil water, mix some salt, put tea leaves into the teapot, boil, and tea-division, drink tea, and clean tea sets. When the Sencha Tea method emerged, those literati, refined scholars, and tea lovers praised it. It gradually matured and became popular, especially in the late Tang Dynasty. Lu Yu founded this Sencha Tea method known as the "Saint of Tea." This method could also be called Lu Yu Sencha Tea Method. The advent of the Sencha Tea method was the embryonic of the Chinese tea ceremony. The technique lasted about 500 years and flourished in the Tang Dynasty, the Five Dynasties, and the Northern and Southern Song Dynasties.

茶禅一味
The Combination of Tea and Zen Buddhism Philosophy

俗话说"吃茶是和尚家风"，僧侣与品茶之风有着极其密切的关系，茶道从一开始萌芽，就与佛教有着千丝万缕的联系，旧时有"自古名寺出名茶"之说，也有说法称茶由野生茶树到人工培植也是始于僧人。

So goes an old saying, drinking tea is a custom among the monks. Monk and tea have a very close relationship from the birth of the tea ceremony.

Ever since there appeared a tea ceremony, there have countless ties between Tea and Zen. Some famous tea comes from those famous temples from ancient times. There was a saying that the artificial cultivation of wild tea also came from monks.

佛教的禅宗认为，参禅时需要有一颗平常心，无妄无欲。茶性平和，香气淡雅含蓄，细品慢啜，回味持久，让人内心宁静，归于平和，这些特性与参禅悟道所秉持的心态有异曲同工之妙，即"禅让僧人有一颗平常心，而茶给茶人以一颗平常心"。日常生活中最平凡不过的"茶"，与佛教中最重要的精神"悟"结合起来，作为禅宗的"悟道"方式，升华出"茶禅一味"至高无上的境界。

The Zen Sect believes that people need to have an ordinary mood without delusion and desire when practicing meditation. You can achieve peace of mind by tasting tea and its gentle taste and its elegant aroma. In this regard, the character of drinking tea has something in common with the attitude of Buddhist meditation. As the ancients said, Monks should have a feeling of security and calmness with Zen, and people who drink tea would have a typical tea attitude. Tea is combined with the essential Buddhism spirit Wu, which means to realize your mind. It can be sublimated to the sovereign power of the combination of Tea and Zen to enlighten Zen.

与文人结缘
Tea and Scholars

　　饮茶能怡神醒脑，有助文思，因此格外得到文人的喜爱，两者结下不解之缘，成为中国人文精神的重要组成部分。于唐朝兴起并得到较大发展的茶文化同时也体现着中国传统文化丰富、高雅、含蓄的特点。

　　Drinking tea can also give you a clear vision and a sense of refreshment for refreshing your literary thought. So the scholars love drinking tea so much and could not live without tea. It became an essential part of Contemporary Chinese Humanity. The tea culture was rising and developing considerably. And it reflected the character of enrichment elegance implicit in the Tang Dynasty. The development of the tea industry is also reflected in Chinese traditional culture.

　　唐朝以来流传下来的茶文、茶诗、茶画、茶歌等，无论从数量还是质量，从形式还是内容，都大大超过了唐以前的任何朝代。饮茶过程既是品味的过程，也是一个自我调节和修养的过程，即灵魂的净化过程。

　　Those tea treatises, poetries, tea paintings, and tea songs have been passed down from the Tang Dynasty in quantity and quality to the level of their form or content. It brought prosperity in the previous dynasties, especially the Tang Dynasty. Drinking tea is the process of tasting and improving self-regulation and accomplishment. Therefore, it is also called soul purification.

茶文化为中华民族异彩纷呈、灿烂辉煌的传统文化增添了新的形式和内涵，注入了旺盛的生命力。饮茶、赋诗、会友，根植于民间百姓的社会生活，为广大人民所普遍接受，逐渐积淀、固定下来，成为一种独具特色的民族文化形态，这是茶文化得以顺利发展，且盛行、繁荣至今的坚实基础。

The tea culture invigorates and energies into China's colorful, diverse, glorious, and splendid culture by building various forms and connotations. Those habits are rooted in ordinary people's lives, such as tea drinking, poem composing, and friendly meeting. People have accepted it, gradually becoming a unique national culture form. It has laid a solid foundation for our tea culture and has flourished.

宋朝茶事
Tea Affairs in the Song Dynasty

遍布街巷
Spreading All Over

经历了唐朝茶业与茶文化启蒙发展阶段，宋朝成为历史上茶饮活动最活跃的时代，除了有内容丰富、技艺高超的"斗茶""分茶""绣茶"等活动以外，民间的饮茶方式则更是丰富多彩。

The Song Dynasty experienced the tea industry and tea culture's enlightenment in Tang Dynasty, spreading, establishment, and development in Tang Dynasty. It had become the most active period of those tea activities in history. Activities such as tea contests, division, and decoration were rich in content and skill. Moreover, tea-drinking ways were also more colorful among the folk.

民间饮茶最为盛行的是在南宋时期的都城临安(今浙江杭州)。

当时繁华的临安城，茶肆经营昼夜不绝，无论是酷暑盛夏还是寒冬腊月，时时有人来提壶买茶。茶肆里面张挂着名人书画，装饰古朴，四季有鲜花装点，前来饮茶的人们络绎不绝，往来如织。

The tea-drinking emerged in Linan's capital during the Southern Song Dynasty, now Hangzhou. Those teahouses were operating round the clock in the most prosperous town Linan then. Whether the weather was blazing hot or bitterly cold in the twelfth month of the lunar calendar, people would pick up a pot to buy tea at every moment. People decorated the teahouses with celebrity calligraphy, paintings, and flowers in all four seasons. The decoration was impressive. A continuous stream of visitors would come to drink.

临安的茶肆通常分成很多种，以适应不同层次的消费者。有一些茶肆，多是士大夫等人与朋友相聚的场所，人们在此不但品茗倾谈，甚至开展体育活动，如蹴球茶坊等。还有作为品茗场所的茶楼、茶馆，其主要顾客多为文雅之士和饱学之人，他们在此把玩乐器，研习弹奏曲目等。还有一些茶馆并非以茶为营生，只是挂名而已，人们在此进行买卖交易，谈事论情，饮酒甚至赌博，成为娱乐场所。

Those various teahouses could meet the different levels of customers' needs. They were predominantly for meeting friends for most literati painters. People enjoyed tea by talking and arranging sports activities like kicking a ball. Some other teahouses were also for those elegant and knowledgeable people to learn the tunes and play instruments. They were multifunctional teahouses; people could talk about business, chat, drink wine, and even gamble. It has tended to be an important place for communication and entertainment.

制茶法
Processing Tea

从朝廷到民间，宋代对茶的品质要求都更为讲究。宋朝历任皇帝几乎皆嗜饮茶，特别是宋徽宗赵佶，虽然不事政务，却在艺术上有着很高的成就，对茶也有深刻的研究，并亲自著成《大观茶论》辑录茶事。他曾不惜重金派人四处寻找新的茶叶品种，大大促进了团茶种类的增多和制茶技术的发展。据《宣和北苑贡茶录》记载，贡茶在宋朝极盛时，有四十余种。

People paid more attention to the tea quality from the palace to the folk in the Song Dynasty. During the Song Dynasty, every emperor loved drinking tea, especially the Emperor Hui of Song Dynasty Zhao Ji. Although he was not good at politics, he still got a high artistic achievements and an in-depth study of tea. He wrote Treatise on Tea to record those tea affairs in person. He invested heavily in sending people to look around for new varieties of solid tea. It accelerated the increase of tea types and the development of technology for processing tea. According to Tribute Tea's record in Bei Yuan Xuanhe, more than forty varieties of tribute tea flourished in the Song Dynasty.

团茶制法比唐朝陆羽在《茶经》中所载的方法又更为精细科学，茶的品质也得到提高。

Because those tea processing methods were more genetic and scientific than the processing way recorded in The Classics of Tea, edited by Lu Yu in the Tang Dynasty, tea quality was also improved.

宋朝末年开始出现散茶制法。到元朝时，团茶已不再流行，散茶则大为发展，"蒸青法"逐渐改为"炒青法"。到了明代，团茶几乎已遭淘汰，炒青散茶则开始大行其道。

People began to process loose tea at the end of the Song Dynasty. But the processing method was no longer fashionable in the Yuan Dynasty. The loose leaves processing method was developed fast and replaced the process of solid tea in the Ming Dynasty. The steamed tea method had gradually changed to stir fixation at that time.

点茶法
Dian Cha Method (Latte Tea Art)

宋朝时期，饮茶方式逐渐发生了新的变化，煎茶法由于繁琐复杂而开始走下坡路，新兴的点茶法成为时尚。蔡襄编著的《茶录》为点茶茶艺奠定了基础。点茶法主要包括备器、选水、取火、候汤和习茶五个环节：在点茶时先将饼茶碾成末，放在碗中待用；烧水时要注意调整炭火；待水初沸时立即离火，冲点碗中的茶末，同时搅拌均匀，茶末上浮，形成粥面，即可饮用。

The way of tea drinking changed gradually during the Song Dynasty. First, the Sencha Method started to turn down due to its complicated and tedious process. Then the new Dian Cha method (latte tea art) became fashionable. The book from Cai Xiang, Tea Record, laid the foundation of the Dian Cha method (latte tea art). The Dian Cha method (latte tea art) mainly included five steps: preparing tea sets, selecting the water, making a fire, soaking, and serving the tea. Grounding those solid tea cakes and putting them aside would be best. Then poke the coals up into a blaze when you boil water. Remove tea soup from heat as soon as the water boils. Pour some water into the bowl to brew tea powders. Then gently mix them to a smooth paste with some floating tea powders. Then it could be drunk.

斗茶的兴起
The Rise of Tea Competition

宋朝时期，随着饮茶的普及，关于茶的活动也日渐丰富起来，民间开始兴起了斗茶的风气，"斗茶"也称"茗战"，用来决定胜负的标准共有两条：一是"汤色"；二是"汤花"。

With the popularity of tea drinking in the Song Dynasty, tea activities were also increasingly rich. Thus there came into being a fashion of tea competition. The tea competition was also called a tea battle. And two criteria for deciding the winner were the tea soup's color and the tea foam on top.

所谓的"汤色"就是指茶汤的颜色，有其固定的标准，即茶汤的颜色以纯白色为最上，其他的颜色则为不正。茶汤纯白色，说明茶叶的采摘、加工都是恰到好处。如果颜色偏青，说明在加工的时候火候不足；如果偏灰，就是过火；如果偏黄，那么则是茶叶的采制环节出了问题。

There was a fixed standard for judging tea soup's color. The best-grade tea soup should be pure white to show perfect picking and processing skill. Others might not have pure color. If the color is partially green, the roasting grade is not enough. Leaves would be partially gray with over-cooked fire. If it was somewhat yellow, the plucking process might be wrong. Tea form was on the surface of the tea soup when pouring them into the tea bowl. The excellent tea foam is well-proportioned.

斗茶中要求水痕出现得越晚越好。要想在斗茶中获胜，就必须把茶末研磨得非常细腻，同时在注水点汤的时候，力道要把握好，不温不火。汤花的最佳效果是汤花出现后久久不散，而且汤花紧紧"咬"住茶盏的边缘，但是绝不能外溢，这就叫做"咬盏"。如果汤花很快散开，或者流溢出来，比赛就会落败。

The water ripple would appear later with better quality. To win the tea competition, you must grind the tea into exquisite powders and control the perfect water-pouring power. The best tea form can be kept long and float close to the bowl's edge but not flow out. The Biting Bowl was called, which means adhering to the tea bowl. You will lose the game if the tea foam is dispersed quickly or spilled over.

分茶的艺术
The Art of Tea Acrobatics

分茶是指饮用末茶时，饮茶人所从事的一种技能性游戏，也叫做"茶百戏"。分茶技艺高超的人可以利用茶碗中的水脉，创造许多绮丽美妙且富于变化的图案来，从图案的变化中得到赏心悦目的乐趣。分茶可以寄托文人的闲情雅兴，培养艺术创作的灵感，体现出人格的品位，是一种精湛的技艺。

Tea drawing was a skill game when people drank dust tea, also known as Cha Bai Xi (Tea Acrobatics). A man with superb drawing skills could use tea soup ripples to create beautiful patterns with various changes and get pleasure from them. Moreover, tea dividing could help maintain the scholars' esthetic mood and inspire spark to embody personality, taste, and consummate skills.

酷爱分茶的蔡襄在《茶录》中提出，要点一盏好茶，首先要严格挑选茶叶。茶以青白色为好，黄白色为差；以自然芬芳者为好，添加香料者为差。其次，为了防止团茶在存放时吸潮而影响品质，在饮用前要进行炙烤以激发其香气。碾罗是冲泡末茶的特殊要求，操作时也要讲究技巧，先用纸将茶裹紧捣碎，然后熟碾并细细筛滤。最后是点汤，要注意控制茶汤与茶末的比例，以及投茶与注水的先后顺序，烧水的温度、茶具的质地颜色以及手法等也有诸多讲究的技巧，如此才能分出一盏美茶。

Cai Xiang had a great love of tea drawing. He pointed out that the tea leaves for diving should be selected strictly. The bluish-white tea soup was the best, and the yellowish-white color was worse. The natural fragrant tea mixed with spice was best. Avoid that the solid tea could be affected by dampness during storage. Tea should be roasted to give out a fragrance. Pan milling was an unusual step for brewing those dust tea. Drinkers should pay attention to their brewing skills. Those tea leaves should be wrapped up and crushed in paper at first. Then filtered and sifted tea soup after roasting. The final step was to pour out tea soup. You should also pay attention to the proportion of tea leaves and tea dust. And then, tea makers should put tea leaves and pour water in their order. The water temperature, the tea leaves, the tea sets' color, brewing skills, and some other process steps also need to be paid attention to and discussed. You could not make good tea without these steps.

图片由茶人章志峰提供
Photos by courtesy of Zhang Zhifeng

宫廷绣茶
Embroidered Tea

宋朝茶文化的发展在很大程度上与宫廷风俗的影响密不可分。因此，无论民间饮茶的文化特色或是形式内容，都带有明显的贵族色彩。茶文化在这种高雅的文化范畴内，得到了全面的发展。宋代贡茶是自蔡襄任福建转运使后，其制作变得更加精良细致，品质上也有了进一步的发展，并由蔡襄亲自研制出了小龙凤团茶。欧阳修评论这种茶为"价值黄金二两"。但是，金可有，茶却不可多得。宋仁宗就格外偏爱饮用这种小龙凤团茶，对其格外珍惜，即使是居功至伟的近臣，也从不随便赐赠，只有在每年的南郊大礼祭天地时，中枢密院的列位大臣才有幸共同分到一小团。大臣们往往舍不得自己饮用，而是用它来孝敬父母或转赠好友。这种茶在赏赐给大臣之前，要先由宫女用金箔剪成龙凤和花草图案贴在上面，因此也被叫做绣茶。"绣茶"是皇廷内的秘玩，由专人掌握此项技术，宫外之人难得一见。

The development of tea culture in the Song Dynasty was related mainly to customs in the palace. So no matter the cultural characteristics or form content, both were with a kind of nobility breathing or excellent manner. Tea culture has been more in-depth and comprehensive development within the range of elegant culture. The masters made the tribute tea elaborate and more detailed in the Song Dynasty. During the Chingli reign (1042–1048 AD), Cai Xiang was Fujian's Officer of Transportation (Zhuanyunshi). He

developed the Dragon and Phoenix tribute tea. Ouyang Xiu commented that this kind of solid tea was worth two gold in price. But the gold was easy to get, and the tea was rare. Emperor Renzong of the Song Dynasty preferred this Dragon and Phoenix solid tea and treasured it. He would not reward this tea freely, even the respected persons with an immeasurable contribution. Some top government ministers could only have the privilege of sharing a solid tea piece when they held the sacrificial rites ceremonies to the gods in the southern suburbs annually. The ministers grudged drinking this tea. They used this tea to honor their parents or make an excellent present for their friends. Before rewarding this tea to the ministers, maids of honor cut the gold foils into Dragon, Phoenix, and Flowers patterns to attach to the solid tea's surface. So it was called tea decoration. Tea decorations were only made in royal courts and mastered by some specially assigned person. It was a rare sight outside the palace.

明朝茶事
Tea Affairs in the Ming Dynasty

由繁及简
From complexity to simplicity

明代是中国茶业与饮茶方式发生重要变革的发展阶段。为去奢靡之风，减轻百姓负担，明太祖朱元璋下令改革茶制，用散茶代替饼茶进贡。伴随着茶叶加工方法的简化，茶的品饮方式也发生了改变，逐渐趋于简化。

The Ming Dynasty was an essential development stage to witness significant changes to drinking tea. The government wanted to reform the institutions to fight extravagance and lessen the tax burden, so the first Ming emperor Zhu Yuanzhang pursued institutional reform. As a result, the loose tea leaves became tribute tea instead of solid tea. Furthermore, with the simplification of tea processing methods, tea drinking had also changed and gradually tended to be simplified.

真正开从简清饮之风的是朱元璋的第十七子朱权。他大胆改革传统饮茶的繁琐程序，并著有《茶谱》一书，书中对茶品、茶具、饮茶方式等茶事活动所涉及的各个方面都提出了明确且具体的要求，特别提出讲求茶"自然本性"和"真味"，对于茶具反对繁复华丽和"雕镂藻饰"，为形成一套从简行事的烹饮方法打下了坚实的基础。

Light drinking was a fashion sparked by Emperor Zhu Quan, the seventeenth son of Emperor Zhu Yuanzhang. Zhu Quan boldly reformed and innovated those traditional drinking ways and wrote the book Tea Spectrum. This book suggests more evident and concrete demand for all aspects of tea activity, such as tea things, tea sets, and tea-drinking methods. In addition, this book puts forward the concept of human nature and pure feelings about tea. He also combated extravagance and waste in decorating tea sets and helped establish a solid foundation for simple tea cooking.

品类增多
Expanding the Tea Product Categories

随着明朝制茶技术的改进，各个茶区出产的名茶品类也日见繁多。宋朝时期闻名天下的散茶寥寥无几，有史料记载的仅有数种。到了明朝，仅黄一正编写的《事物绀珠》一书中收录的名茶就有近百种之多，且绝大多数属于散茶。

在明清时期，茶叶的制作形式得到了真正的飞跃和发展，黑茶、青茶、红茶、花茶等各种茶类相继出现和扩大。青茶，即乌龙茶，是明清时期由福建首先制作出来的一种半发酵茶类。红茶最早见之于明朝中期刘基编写的《多能鄙事》一书。清朝时，随着茶叶贸易的发展，红茶从福建很快被传播到云南、四川、湖南、湖北、江西、浙江、安徽等省。此外，在各地茶区还出现了工夫小种、紫毫、白毫、漳芽、清香和兰香等许多名优茶品，极大地丰富了茶叶种类，推动了茶业的发展。

The famous tea category increased daily in every tea-producing area with the technological advancement of tea processing in the Ming Dynasty. As a result, there were fewer kinds of renowned tea, according to historical records. Until the Ming Dynasty, nearly 100 kinds of famous teas were recorded in the book of Shi Wu Zu Zhu, edited by Huang Yizheng. Most of those were piece-tea. The variety of Chinese tea underwent a historical leap and development in the Ming and Qing Dynasties with the emergence and broadening of Dark Tea, Oolong Tea, Black Tea, Scented Tea, and other teas. Celadon tea is also called Oolong Tea, a semi-fermented tea.

People created the Oolong Tea in Fujian Province in the Ming and Qing Dynasties. The Black Tea was first documented in Duo Neng Bi Shi in the Ming Dynasty, edited by Liu Ji. With the tea trade development, Black Tea spread quickly from Fujian Province to Yunnan, Sichuan, Hunan, Hubei, Jiangxi, Zhejiang, Anhui, and other provinces. Furthermore, Gongfu Xiaozhong Black Tea, Purple Tip Tea, White Tip Tea, Zhangzhou Tip Tea, Xuan Tip Tea, Light Aroma Tea, Orchid Aroma Tea, and other famous tea varieties have appeared in each tea-producing area. They greatly enriched tea varieties and promoted the development of the tea industry.

泡茶法
Brewing Tea

泡茶法是将茶放在茶壶或茶盏之中，以沸水冲泡后直接饮用的便捷方法。唐及五代时期的饮茶方式都以煎茶法为主，宋元时期以点茶法为主，泡茶法虽然在唐代时已经出现，但是始终没有被传播开来，直到明清时期才开始流行，并逐渐取代煎茶法和点茶法而成为主流。

The most economical and convenient tea brewing method is to put the tea leaves into a teapot or tea bowl and then pour them into the water for drinking. The Sencha Tea was the primary method in the Tang Dynasty and the Five Dynasties. And the standard way in the Song Dynasty and Yuan Dynasty was the powder tea method. Though the tea brewing method had already been presented in the Tang Dynasty, this method was spread in the Ming and Qing Dynasties and gradually replaced the loose-leaf tea method. So the Sencha Tea Method became more mainstream at that time.

明清时期的泡茶法使用较为普遍的是用壶冲泡，即先置茶于茶壶中冲泡，然后再分到茶杯中饮用。据古代茶书的记载，壶泡法有一套完整的程度，主要包括备器、择水、取火、候汤、投茶、冲泡、酾茶、品茶等。泡茶之道孕育于元末明初时期，正式形成于明朝后期，到清中期之前发展到鼎盛阶段并流传至今。今日流行于福建、两广、台湾等地区的"功夫茶"即是以明清的壶泡法为基础发展起来的。

People generally brew tea in a teapot in the Ming and Qing Dynasties. Firstly, brew some tea leaves with water into the teapot and divide the tea soup into teacups. According to the ancient historical record, Tea-brewing in a pot was with a set of complete procedures. It mainly included: preparing tea sets, selecting water, making a fire, soaking, putting tea leaves, brewing, pouring out the tea soup, and tasting tea. Those brewing ways were conceived during the late Yuan and early Ming Dynasty and originated in the late Ming Dynasty. It flourished in the middle of the Qing Dynasty and has passed so far. The Gongfu Tea was developed with the teapot brewing method in the Ming and Qing Dynasties, popular in Fujian, Guangdong, Guangxi, and Taiwan.

清朝茶事
Tea Affairs in the Qing Dynasty

茶叶的生产
Tea Production

据古籍史料显示，明清时期在前朝的基础上出现了很多新的茶树种植和茶叶生产加工技术，对于茶树生长规律和特性的掌握也有很大进步。在清朝的福建北部一带，茶农们对一些珍稀名贵的优良茶树品种还开始采用了压条繁殖的方法。在茶园管理方面，明清时期在种植上有了关于灌溉施肥等更加精细的要求，在抑制杂草生长和茶树与其他植物间种方面也有精辟见解。此外，明清时期在茶叶采摘技术方面较前朝也有了较大的提高和发展。

According to historical data and some ancient books, many planting and tea processing technologies advanced breathlessly during the Ming and Qing Dynasties. It was also significant progress. Some tea farmers had already used propagation technology to plant rare and precious tea varieties.

More detailed irrigation and fertilization requirements were in the Ming and Qing Dynasties' tea garden management. It also had some incisive views on restraining the growth of weeds. Furthermore, tea plucking technology was also highly improved at that time than before.

从调饮到清饮
From Flavoring Drinking to Light Drinking

调饮法与清饮法有着显著的区别，各有优势。纵观饮茶历史，其发展的顺序是由调饮法逐渐过渡到清饮法。在饮茶之风兴盛的唐代，人们在饮茶时普遍以佐料调味。到了现代，只有部分边疆民族地区还继续沿用调饮的方式，而清饮法早已得到普及。

所谓调饮法，即在茶汤中加入糖或盐等调味品，以及牛奶、蜂蜜、果酱、干果等配料，调和后一同饮用。调饮法因地区和民族的不同而呈现出复杂多样的特点，其中最具代表性的咸味调饮法有西藏的酥油茶和内蒙古、新疆的奶茶等；甜味调饮法有宁夏的"三泡台"；调味可咸可甜的饮茶法有居住在四川、云南一带山区民族的擂茶、打油茶等。而清饮法就是不加入任何调料，饮用单纯的茶汤，来品尝真正的茶味。时至今日，我国大部分地区的人们仍采用此种饮法。

There was a marked difference between flavoring tea drinking and light drinking. Each method had its specific advantages. Then the development was from flavoring drinking to light drinking. The flavoring tea-drinking way presented a complex and diverse look by the different nationalities and ethnicities. The most representative salt-flavoring tea was mixed with Tibet's butter tea and Xinjiang and Inner Mongolia-style milk tea. Some teas mixed with sweet flavors include San Pao Tai method in Ningxia Hui Autonomous Region. Those ethnic minorities who live in the Sichuan and Yunnan Province mountains prefer flavored tea soup with salt and the sweet taste of Lei Cha and Oily Tea. Light drinking was a way without any flavoring to experience the real taste of pure tea. The vast majority of the Han nationality in China still uses this method to drink tea.

普洱贡茶
Pu Er Tribute Tea

普洱茶是茶中珍品，在清代不但深受民间百姓的喜爱，还被上贡朝廷，供皇族大臣们品饮，甚至将其作为珍贵礼品馈赠他国。普洱茶茶味浓醇，具有性温味香、消积去腻等诸多利于人体的保健作用，这些特点正适合游牧出身，以肉食为主的清廷满族皇亲国戚的需要。清朝政府规定每年茶农须上缴普洱贡茶，由地方官吏负责组织运送。在进贡清宫的普洱茶中，主要有来自云南西双版纳原始森林的大叶种极品"金瓜贡茶"，还有其他各地进贡的小叶种茶，其中的"女儿茶"、团茶、茶膏等深得王公贵族的喜爱。一时间，宫中饮普洱茶之风成为时尚，既有清饮，也有用来熬煮奶茶。朝廷之风得到民间的大力效仿，普洱茶在清朝声名大振，流传甚广。

Pu Er Tea is one kind of precious tea. It has become popular among folks in China and was the tribute for royal and imperial courts for tasting. It was even regarded as a special gift to other countries. The taste of Pu Er Tea is strong and mellow. It's warm and with an enjoyable sweet aroma. Lots of healthcare functions of Pu Er Tea were validated. It can help people digest greasy food. These characteristics were just right to fit the Manchus Royal Families' needs from the ethnic minority communities and those who preferred cattle and sheep-based diets in the Qing Dynasty. The Qing Dynasty government set a minimum of 33,000 kilograms to be paid tribute from tea farmers and transported by the native officials. Those Pu Er teas were mainly the big leaf species, such as Jin Gua Tribute Tea from the primordial forest in Xishuangbanna of Yunnan Province, and other small leaf species which were tributed from different places in the country. The Nver Tea, Solid-tea, and Tea Cream got love and praise from the ordinary to the royal. There was a moment of drinking Pu Er Tea became a fashion in the palace. Light drinking and boiling milk tea with Pu Er Tea was fashionable. This fashion was widely replicated and expanded in the folks. The reputation of Pu Er Tea was significantly boosted and spread widely.

茶馆兴盛
Teahouse Prosperous

明清之际，茶馆开始兴盛。特别是清代，各种茶馆、茶肆、茶档作为百姓日常重要的活动场所，如雨后春笋般迅速发展起来。人们在此既可饮茶，也可会友。书生吟诗作对，商人高谈阔论。据史料记载，到清朝末期，仅皇都北京城内有规模的茶馆就达数十家。即使在乡野之间，茶馆的发展也不亚于繁华都市，特别是江南的苏浙一带，有的小镇虽只有居民数千家，可是茶馆却有上百家之多。

The teahouse gradually flourished in the Ming and Qing Dynasties as an essential place for holding ordinary people's daily activities. Different teahouses, teashops, and tea stalls had rapid development. People could drink tea and meet friends, recite poems, and talk eloquently at a teahouse. According to the historical record, dozens of tea houses in the Imperial Capital, Beijing City, and as many as 66 tea houses in Shanghai. The development of the teahouse in the countryside was not second to a busy modern city. There were hundreds of teahouses in a small town where

only thousands of citizen families lived, mainly in Jiangsu and Zhejiang Provinces.

清朝的茶馆依据经营内容和功能特色的不同，主要有以下几种：品茗饮茶之地；饮茶兼饮食之地；还有最富特色的听书赏戏之地。除此之外，在江南乡镇，有的茶馆还兼做赌博场所，有时也充当排解百姓纠纷的仲裁场所。

There are several significant teahouse types during the Qing Dynasty. The most entertaining style is to drink tea with talking stories according to different operation modes and functional features. Besides these, in Jiangnan towns, some tea houses were also places for gambling and arbitration to ease the worry.

地方茶俗的发展流传
The Development and Spread of Local Tea Custom

清代茶文化在民间的深入,还突出表现在一些地方茶俗的发展流传,形成了各具特色的地方茶文化,例如茶叶的生产习俗、茶业经营、日常饮茶、以茶待客、节日饮茶、婚恋用茶、祭祀供茶、茶馆文化、茶事茶规等,涉及各个方面,内容丰富多彩。

It makes an endless effort to put tea's cultural depth into people's minds during the Qing Dynasty. Those developments and spread are also manifest in some places. It made its own regional tea culture with distinctive characteristics: production customs, industry management, daily drinking, treating guests customs, festival drinking customs, wedding tea, sacrifice tea, tea house culture, tea activities, and tea rules. Those were expressed by rich and colorful imagination in all aspects.

地区不同,风俗不同,饮茶也不同。北方人喜欢喝花茶,江浙人喜欢喝绿茶,福建人擅长饮乌龙茶,两广地区则喜红茶,而边疆民族饮用的"边茶"属于黑茶。

As for the different regions and customs, the tea-drinking habit was also not the same. Most northerners prefer Scented Tea. However, in the Jiangsu and Zhejiang Provinces, people prefer Green Tea. Fujian people like to drink Oolong Tea. Guangdong and Guangxi people prefer Black Tea. The frontier ethnic region that consumes the border tea is Dark Tea.

同时，明清时期还涌现出大量悦耳动听的茶歌、别开生面的茶舞、幽默风趣的茶戏和曲折动人的故事，可谓各种与茶相关的文化艺术百花齐放，繁华似锦。

And many melodic songs, characteristic tea dances, humorous tea dramas, and moving stories about tea emerged during the Ming and Qing Dynasties. All kinds of tea-related arts and culture were blooming like a hundred flowers.

茶类篇

The Category of Tea

茶者，南方之嘉木也。一尺、二尺乃至数十尺；其巴山峡川有两人合抱者，伐而掇之。其树如瓜芦，叶如栀子，花如白蔷薇……

The tea tree is a precious tree species in southern China. The height of the trees varies from one or two feet to tens of feet. Trees grow around the Bashan and Xiachuan areas. (Currently Eastern Sichuan Province and Southwest Hubei Province). Some trees are so large that two adults can hold them together. You have to cut branches off the trees to pick up the bud leaves…

籯：也叫篮、笼或筥。用竹子编织而成，容积通常为五升，也有一斗、二斗或三斗的，茶农采茶时将其背在肩上用于盛茶。

The tool Ying has been widely used for holding and carrying tea leaves since thousand years ago. It was also called Lan Zi (basket), Long Zi (cage), or Ju (bamboo basket). Those tools were made of bamboo with five liters and three different volume types: a pipe, two, and three. Tea farmers placed the tool on their shoulders when they plucked the tea leaves.

灶：生火用的灶，不要使用带烟囱的。

Zao (stove). Please do not use the stove with a chimney to build fires because it can converge to the focal point and destroy those leaves.

釜：用"唇口"者。指使用带有边沿的容器。

Fu (kettle). A tool Fu (kettle) is a covered container for boiling water. It has a handle and a spout for the water to come from the kettle.

甑：是用木头或陶土制成，腰部用泥封住的容器。内有蒸屉，并用细竹片牢系。开始蒸茶的时候，将芽叶放在蒸屉里，蒸熟后即可取出。蒸干时可往甑中加水，然后用分叉的枝条翻动摊晾蒸好的芽叶，以防止茶汁的流失。

Zeng was an ancient earthen utensil for steaming rice. Its material was wood or pottery clay. It must be sealed up with mud at its narrowed waist portion. Put a bamboo grate to separate the water. Then put the tea leaves inside the steaming drawer when they started to steam leaves. The pot is bubbling on the fire. Then pour the tea soup out after cooking. Add boiled water into the tool Zeng when the teapot on the fire is nearly dry. To avoid leaf juice loss, stir tea leaves and bamboo shoot buds with a three-pronged wooden stick.

忤臼：又叫碓，日常使用的即可。

The tool Chu Jiu was also called Dui. It is better for frequently used.

规：又叫模或棬，通常用铁打制而成，呈圆形、方形或各种其他形状，是压制饼茶的模型。

The tool Gui was also called Mo or Quan. It was usually made of iron. The shapes of Gui are circular, square, and varied.

承：又叫台或砧，用石料制成。也可将槐木或桑木埋进土中，露出半截，使其牢固而不易晃动即可。

The tool Cheng was also called Tai or Zhen. Its material was stone. The sophora japonica and mulberry wood were buried in the soil and unable to be shacked out.

檐：又叫衣。通常用油绢或雨衣、单衣等做成。将"檐"放在"承"上，再将"规"放置在"檐"上，即可用来压制饼茶。做好后，拿开压制好的饼茶即可继续制作下一块。

Yan, also known as Yi, can be made of oily silk, worn-out raincoats, or single clothes. To make compressed cake tea, place "yan" on top

of "cheng" and then place a model on top of "yan." After pressing it together, pick it up and make the next one.

芘莉：又叫做籯子，用两根长约1米的竹竿制成，身长约85厘米，柄长约15厘米。中间用篾编织成类似筛箩的形状，直径约70厘米，用来铺放茶叶。

The tool Bi li was also called Ying Zi or Peng Lang. It was a bamboo tray or bamboo basket. It was made of two bamboo strips about three feet long. And people weaved with bamboo split to be two feet five inches long and two feet wide. Like a bamboo basket, this handle was about five inches and fifteen centimeters. People use it to spread out tea leaves.

棨：又叫锥刀。手柄用坚硬的木材制成，用来给饼茶穿洞。

The tool Qi was also called a conical knife. Its material is solid wood for making a perforation in solid tea.

扑：又叫鞭。用竹子编成，用来把饼茶串起来。

The tool Pu was also called a whip. It was made of bamboo to string those solid tea cakes for carrying.

茶的分类
Classification of Tea

  中国作为茶叶的故乡，种茶、品茶历史悠久，产茶量极为丰富。因疆土广袤，各地环境气候不尽相同，茶的种类繁多，千差万别。茶树的生长习性分为各式各样，摘下来的鲜茶经过不同的加工方式又产生不同的特性。目前茶的分类有很多标准，尚未统一，根据不同的标准产生了不同的分类方法。不同类型的茶从外形、色泽、香气、滋味、功效方面相比较又各有千秋……

  As the hometown of tea, China has rich resources and a long history of the planting and tasting tea. Chinese tea varies widely with its vast territory and different climatic conditions. There are a variety of growth habits of tea trees. Those fresh leaves were created with unique characters with varying methods of processing. There are many standards for tea classification which are not unified. Other tea types can be compared in every aspect, such as appearance, color, aroma, taste, and efficacy. But each method has its advantages...

  根据生长地特有的文化背景，综合茶叶本身的特性，很多中国名茶被赋予了各色雅致的名称，令人只看茶名便已对其产生好感，亟待享受品饮的乐趣。深入了解其背后蕴涵的丰富文化内涵，也已成为中国茶文化中一道别致的亮丽的风景。

Many famous Chinese teas were given elegant names according to their unique cultural identity of growing place and characteristics. People would like to enjoy tea when they hear their beautiful words. A deep understanding of rich cultural connotations has been a unique bright scenery of Chinese tea culture.

按加工方法分类
Classification by Processing Method

茶的分类方法繁多，目前使用最为广泛的方法之一是根据制作茶叶时的工艺不同来划分。这主要是指在制茶过程中是否有发酵这一步骤。

There are various classification methods. At present, the most widely used method is the processing method of classification, mainly called the fermenting degree.

绿茶，在制造过程中没有发酵工序，茶树的鲜叶采摘后经过高温杀青，去除其中的氧化酶，然后经过揉捻、干燥制成。成品干茶保持了鲜叶内的天然物质成分，茶汤青翠碧绿。

Green Tea is not fermented in the process. Those fresh leaves were under stir fixation with high temperature to decrease oxidase activities until the end. Then tea leaves could be made after rolling and drying. We should maintain its native inclusions of fresh leaves.. The tea soup is fresh and green.

青茶（乌龙茶）、白茶、黄茶等为部分发酵茶，制作时较绿茶相比多了萎凋和发酵的步骤，鲜叶中一部分天然成分会因酵素作用而发生变化，产生特殊的香气及滋味，冲泡后的茶汤色泽呈金黄色或琥珀色。

Oolong Tea, White Tea, and Yellow Tea are partially fermented tea. There are two more processing steps than Green Tea. The unique step of withering and fermenting are processed in these teas. Some of the natural components of fresh leaves are changed by enzymes which producing and creating unique fragrances and tastes. The color of the tea soup is golden and amber.

红茶为全发酵茶，制作时萎凋的程度最完全，鲜茶内原有的一些多酚类化合物氧化聚合生成茶黄质和茶红质等有色物质。其干茶色泽和冲泡的茶汤以红、黄色为主调。

Black Tea is a fully fermented tea with the best and complete degree. Some original polyphenol compounds would produce theaflavins and thearubigins by chemical oxidation polymerization. The tea leaves and tea soup color are mainly red and yellow. Dark Tea is a post-fermented tea.

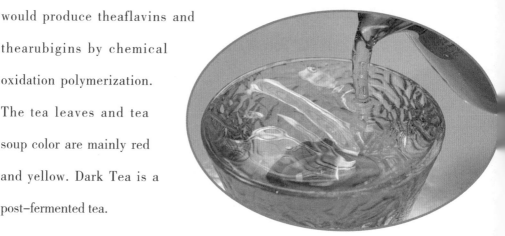

按萎凋与不萎凋分类
Classification by Withering

茶鲜叶采摘下来后，首先要放在空气中蒸发掉一部分水分，这个过程称之为萎凋。按茶叶制作过程中是否需要进行萎凋来划分，茶的种类可分为萎凋茶和不萎凋茶。

Those fresh leaves were exposed to air to evaporate some of their moisture. A carding to whether the tea needs to be withering during the production process, the kinds of tea can be divided into withered tea and imwithered tea.

按茶的季节性分类
Classification by Seasonal Factors

在中国绝大部分产茶地区，茶叶的生长和采制是分季节性的。按照季节变化，可将茶叶划分为春、夏、秋、冬四季茶。

Tea is a kind of seasonal crop in most tea-producing areas of China. They can be divided into four types: spring tea, summer tea, autumn tea, and winter tea on the seasonal variation.

春茶为3月上旬至5月上旬之间采制的茶，采摘期约20～40天，

随各地气候而异。由于春季气温、降雨量适中，无病虫危害，春茶茶叶鲜嫩，香气馥郁，品质最佳。

Those farmers would pluck the spring tea leaves from early March to early May. It will last about 20 to 40 days and differ in climate. The moderate tem perature and rainfall in spring,no pests and diseases make the spring tea leaves fresh, fragrant and excellent qualith.

夏茶在夏至前后采摘，一般为5月中下旬到6月末，是春茶采摘一段时间后所新发的茶叶。夏茶茶叶的新梢生长迅速，不过很容易老化。由于受高温影响，茶叶中的氨基酸、维生素的含量较春茶明显减少，味道也比较苦涩。

The summer tea can be picked up during the summer solstice from early March to early June. Tea tree sprouts new leaves for plucking after spring tea. These newborn tea shoots of summer tea proliferate but are liable to deteriorate. The high temperature reduced the amino acids and vitamin content more than the spring tea. The taste is slightly bitter.

秋茶为7月后采摘的茶叶。秋高气爽，有利于茶叶芳香物质的合成与积累，所以秋茶具有季节性高香。

The autumn tea could be plucked up after July. The autumn sky is clear, and the air is crisp. The climate promotes the growth and accumulation of aromatic substances in tea. So autumn tea has a high seasonal flavor.

冬茶为秋分之后所采制之茶,因气候寒冷,中国大部分地区均不产冬茶。只有海南、福建和台湾因气候较为温暖,尚有生产。

The winter tea could be plucked up after the autumnal equinox. However, most Chinese farmers do not produce winter tea because of cold weather. There are only some warm areas, such as Hainan, Fujian and Taiwan, which still make winter tea.

按茶的生长环境分类
Classification according to the Growing Environment

根据茶树生长的地理条件,茶叶可分为平地茶、高山茶和有机茶几个类型,品质也有所不同。平地茶因茶树生长速度较快,导致茶叶较小,叶片单薄;加工之后的茶叶则条索轻细,香味比较淡,回味短。

相比平地茶,高山茶可谓得天独厚。茶树一向喜温湿、喜阴,而海拔比较高的山地正好满足了这样的条件,也就是人们平常所说的"高山出好茶"。

Tea can be divided into plain tea, high mountain tea (alpine tea), and organic tea. The quality is also different. Plain tea is much more familiar, and it grows quickly. But that also means that these tea leaves are small and thin. The tea leaves after processing are clear, with a light aroma and short aftertaste. Compared with the plain tea, the high mountain tea is unique. The tea trees is sciophilous and hydrophilous. So those high-

altitude mountainous regions can meet the conditions. It is said that we usually talk about it, which is good teas come out from the mountains.

温润的气温，丰沛的降水量，浓郁的湿度，以及略带酸性的土壤，促使高山茶芽肥叶壮，色绿茸多。制成之后的茶叶条索紧结，白毫显露，香气浓郁，耐于冲泡。

Under the mild temperature condition with plentiful precipitation, the slightly acidic soil makes the mountain tea sturdy and bold. It is green with plenty of fine fuzz. The tea leaf is tight with a white tippy and heavy flavor. This tea can endure repeated infusions.

有机茶是近年来出现的一个茶叶新品类，或者说是一个新茶叶鉴定标准。有机茶是指在完全没有污染的环境中，种植生长出来的茶芽，又在严格清洁的生产体系里生产加工，并且遵循着无污染的储存和运输要求，且要经过食品认证机构审查和认可。

Organic Tea is a new category of Tea. It has emerged in recent years. Or it is the unique identification criterion of Tea. Organic Tea grows in an unspoiled environment. Leaves are processed under a strict and clean production system. This whole production followed the general request for pollution-free storage and transportation. It was examined and approved by a specific certificate authority of food.

按茶的品质特点分类
Classification according to the Quality Characteristics

根据加工方法及品质特色的不同，茶可分为六大类，即绿茶、红茶、青茶、黄茶、白茶和黑茶，这也是传统茶文化中最常使用和最为人们所熟知的分类方法。

According to the different processing methods and quality characteristics, Chinese tea has six main categories: Green Tea, Black Tea, Oolong Tea, Yellow Tea, White Tea, and Dark Tea. It is also one of the most commonly used and best-known classification methods.

绿茶的制作没有经过发酵，较多地保留了鲜叶内的天然物质，因此成品茶的色泽、冲泡后的茶汤和叶底均以绿色为主调。同时，由于营养物质损失少，绿茶也被视为更益于人体健康的茶。绿茶是中国种类最多、产量最大的茶类，此外，绿茶也是生产花茶的主要原料。

Green Tea production is without fermentation. It can retain most natural substances, so the tea soup and leaves are still green. With less loss of nutrients, Green Tea is also considered as the healthier tea. Green Tea is the tea with the most types and has the most massive production in China. Furthermore, Green Tea is the primary material of Scented Tea.

红茶，属于发酵茶，因其干茶色泽、冲泡后的茶汤和叶底以红

色为主调而得名。红茶的香气最为浓郁高扬，滋味香甜醇和，饮用方式多样，是全世界饮用国家和人数最多的茶类。

Black Tea is a fully fermented tea. It is famous for the red color of dry tea leaves. The aroma is high and sharp, with a sweet and mellow taste. A variety of drinking methods can be applied to Black Tea. Black Tea has already been tasted by the most significant number of countries and the most massive number of people.

青茶，主要指乌龙茶，属于部分发酵茶，融合绿茶和红茶的清新、芬芳、甘鲜于一身，品质极为出众，得到很多海内外茶人的喜爱和追捧。

Oolong Tea is a partially fermented tea with an oxidation time between Black and Green Tea. As a result, it has fresh, fragrant, and sweet between Green Tea and Black Tea. This tea is exceptionally outstanding and favored by lots of people worldwide.

黄茶的黄色来自制茶过程中的闷黄，独特的制作工艺使其冲泡后呈现出"黄叶黄汤"的特色，且毫香浓显，滋味鲜醇。

The color of the yellow leaves is made in the process of sweltering. The unique technique creates yellow tea leaves and tea soup features with a strong pekoe flavor and fresh and mellow taste.

白茶采用叶表有白色茸毛的细嫩芽叶制成，制作工程中不揉不炒，完整地保留了其原有的外表。优质成品茶毫色银白闪亮，滋味清新甘爽，是不可多得的珍品。

The White Tea is made from those tender and white fuzz tea leaves, then processed without rolling and firing. The original appearance of leaves has been preserved with silvery-white pekoes. The taste of high-quality White Tea is fresh and sweet. It is a rare treasure.

其他分类方法
Other Classification Methods

长期以来，民间习惯利用制茶工艺中的焙火程度来界定茶叶，根据焙火的轻重将茶叶分为生茶与熟茶两种。又根据火候的轻重，将其中的熟茶分为轻火茶、中火茶和重火茶三种。

Chinese people were getting used to the classification according to the degree of roasting. The different degree of roasting divides tea into unfermented and fermented. The fermented tea is also divided into light baked, regularly baked, and strongly baked.

各种茶因制造技术及采摘部位的差别而呈现不同的外观，常见的有条形茶、半球形茶、球形茶、扁形茶、碎形茶、针形茶、片形茶、圆形茶、雀舌形茶等。同种类茶的茶青因市场的供需，可依不同制造方法制成各种不同外观的茶叶。按茶叶成品的聚合形态，茶叶种类又可分为叶茶、砖茶、末茶等。

The various types of tea could present different appearances with different process methods and other plucking parts. The most common types are loose tea leaves, hemispherical tea, spherical tea, flat tea, broken tea, needle-shaped tea, slice tea, round tea, sparrow tongue tea, etc. Those same kinds of tea leaves can be created in different shapes and sizes according to the market supply and demand. The different shapes and tightness indicate tea can also be divided into leaf, brick, and fanning.

还有一种较为通用的分类方法是将中国茶叶分为基本茶类和再加工茶类。基本茶类分为六类，即绿茶、黄茶、黑茶、白茶、青茶、红茶。以这些基本茶类作原料进行再加工后的产品统称为再加工茶类，主要有花茶、紧压茶、萃取茶、果味茶、药用保健茶和含茶饮料等。

There is one more common classification method. Chinese tea is divided into significant types of processed tea and reprocessed tea. The major tea types include Green Tea, Yellow Tea, Dark Tea, White Tea, Oolong Tea, and Black Tea. The materials of reprocessed tea are those significant tea types. There are some primary majors: scented tea, compressed tea, extracting tea, flavored tea, medicinal health protection tea, and tea drinks.

绿茶
Green Tea

顾名思义，绿茶以汤色碧绿清澈，茶汤中绿叶飘逸沉浮的姿态最为著名，它滋味鲜爽，回味无穷，品之神清气爽。绿茶中的天然物质保留较多，儿茶素是绿茶成分中的精髓部分，故绿茶的滋味收敛性强，对预防衰老、抗癌、杀菌、消炎，甚至降脂减肥等均有特殊效果，为其他茶类所不及。饮绿茶不但是精神上的享受，更能保健防病，有益身心。工作繁忙的都市白领喝上一杯绿茶，可以有效地缓解疲劳。夏天饮用，更能消暑解热。

As the name suggests, Green Tea is famous for its brilliant tea soup and elegant tea leaves in the tea soup. The taste is fresh and brisk, with an endless aftertaste. Most natural substances are retained. Tea catechin is a part of the essence of Green Tea. So the astringency taste of Green Tea is strong. Green Tea has special effects on anti-aging, anti-cancer, pactericidal, anti-inflammatorg, even reducing fat and weight loss. Green Tea is better than other

teas on this point. Drinking Green Tea not only is a spiritual experience but also can promote health and disease prevention. It is good for your health. The busy white-collar workers in urban can relieve fatigue effectively and relax by drinking Green Tea. It is also a good drink for relieving heat and sunstroke.

绿茶的品质
Quality of Green Tea

炒青绿茶
Stir Fixed Green Tea

加工过程中采用炒制的方法来干燥的绿茶被称为炒青绿茶。由于炒制过程中手法变换及机械外力的影响，使得成品茶叶呈现出长条形、圆柱形、扇形、针形、螺形等不同的形状，故又可分为长炒青、圆炒青、扁炒青等。

As the roasting method was used during the drying process, some type is called roasted Green Tea. With the changeable technique and mechanical force, tea leaves are made into strips, cylindrical, fan-shaped, needle-shaped, spiral, etc, so it can be divided into long Pan-Fried Green, round Pan-Fried Green. flat Pan-Fried Green and so on.

烘青绿茶
Hot Air Fixation Green Tea

用烘笼烘干进行干燥的绿茶为烘青绿茶，初制工序分为杀青、揉捻、干燥三个过程。烘干后的毛茶再经精加工后大部分用作熏制花茶的茶坯，加入鲜花，利用茶叶的吸附性，待到鲜花吐出香味，合理搅拌和窨制，形成既融入花香又保持了茶香的成品花茶。如今，部分名优绿茶也采用烘青制法制作。

Hot Air Fixation Green Tea is dried with a constant hot wind from the bamboo-frame dryer. The primary process of Green Tea can be divided into three parts fixation, rolling, and drying. Those dried raw tea leaves would be used as the tea dhool for scenting the flower tea in the exact processing and then added fresh flowers, using the adsorbabihty of tea until the flowers spit out the fragrance, reason-able stirring and scenting. Now some kinds of famous Green Teas are made by hot air fixation.

烘青绿茶外形完整稍弯曲，锋苗显露，色泽墨绿，香清味醇，汤色明亮，但是香气一般不如炒青绿茶高。烘青绿茶根据外形分为条形茶、尖形茶、片形茶、针形茶等。一些烘青名茶品质特优，特种烘青主要有黄山毛峰、六安瓜片、天山绿茶、峨眉毛峰等名茶。

The hot air fixation Green Tea bends slightly and has a black-green color. Its aroma is fresh and clean. The taste is mellow. The tea soup is brilliant. But its aroma is less strong than roasted Green Tea. According to its shape appearance, stir fixation Green Tea can be divided into loose tea, acuminate tea, flake tea, flat tea, broken tea, needle-shaped tea, slice tea, round tea, and sparrow tongue tea. The quality of some hot air fixation Green Teas is excellent. There are some significant types: Yellow Mountain Fuzz Tip, Luan Guapian, Mountain Tianshan Green Tea and Emei Mountain Mao Feng, and so on.

晒青绿茶
Sun-Fixed Green Tea

　　晒青绿茶是绿茶里一个比较独特的品种，鲜叶经过锅炒杀青、机械揉捻之后，不再采用人工加工，而是直接进行日照。因太阳光的照射温度比较低，所以晒青所需要的时间也比较长。在这个过程中因没有非自然因素的破坏，所以最大程度地保留了茶叶内的天然物质，使得成茶滋味浓厚，并且有一种馥郁的青草味，甚至还可以品尝出"浓浓的太阳味"。不过晒青绿茶往往不宜直接饮用，而是用来制作紧压茶，比如砖茶、沱茶、普洱生茶等，有效地延长了它的保存时间。

　　Sun-fixed Green Tea is a unique variety. The sun evaporates the fresh leaf's moisture instead of the manual work. Because the temperature is low, this step might take more time. During this process, it can maximum retain those natural tea substances to keep the rich flavor taste of tea soup without any damage by non-natural factors. It also has a strong taste and rich grassy smell. You can even taste the strong smell of the sunshine. But we always do not drink Sun-fixed Green Tea directly. Instead, we use it as the material for making compressed tea, such as Brick tea, Tuo tea (bowl-shaped compressed mass of tea leaves), Pu Er tea, and so on. The optimal preservation time could be extended effectively.

根据产地不同，晒青绿茶可分为滇青、川青、陕青等品种，其中以云南大叶的滇青品质为最佳，可作为沱茶和普洱生茶的原料。其他各地区的晒青茶虽也各有千秋，但均不及滇青。

Depending on the place of origin, Sun-fixed Green Tea can be divided into several types, Dian Qing tea, produced in the Yunnan Province, and Chuan Qing Tea, made in the Sichuan Province—Shan Xi province, Shan Qing Tea, etc. The quality of Dian Qing Tea is the best among these. They are also the material of Tuo tea (bowl tea) and Pu Er tea. Those Sun-fixed Green Teas also have good points from any other region. But they are not as good as Dian Qing Tea.

蒸青绿茶
Steam Fixed Green Tea

利用高温蒸气将茶树鲜叶杀青所制成的绿茶被称为蒸青绿茶。由于蒸气破坏了鲜叶中酶的活性，形成干茶色泽深绿、茶汤浅绿和茶底青绿，即"三绿"的品质特征，茶汤颜色清澈，十分悦目，但茶香较闷，带青气，涩味也较重，不够鲜爽。蒸青绿茶自唐朝时传入日本，启发了日本茶道。至今，日本茶道所用的茶仍是蒸青绿茶。

Steam Fixed Green Tea is known for its unique process. Those fresh tea leaves were processed with steam fixation at high temperatures. High-temperature steam destroys relevant enzyme activities to make the deep green dry tea, tea soup tender green, and blueish-green brewed leaves. The characteristics of these three greens were that the color of tea soup is clearly a comfortable eye-feast. But it also has a sulk green odor with strong astringency and is not enough fresh and sweet. The tea used in the Japanese tea ceremony is the steaming Green Tea introduced into Japan from China during the Tang Dynasty.

相比日本,中国的茶叶制作有了很大的改变,蒸青绿茶不再普及。虽然湖北的恩施、当阳,江苏的宜兴至今还在生产蒸青绿茶,但基本上采用的也是日本工艺,产品也以返销日本为主。

But compared with Japan, Chinese tea processing has changed dramatically over the years. For example, steaming Green Tea is not popular any longer. However, it is produced in Enshi and Dangyang in Hubei Province and Yixing in Jiangsu Province, mainly with Japanese craftsmanship and manufacture for re-exporting back to Japan.

名茶种类
Variety of Chinese Famous Green Tea

西湖龙井是绿茶中最受欢迎的品种之一，产于浙江省杭州市西湖群山之中。龙井茶形光扁平直，状如雀舌，色翠略黄，滋味甘鲜醇和，香气优雅清高，汤色碧绿，叶底嫩匀成朵。

West Lake Longjing Green Tea is one of the most popular varieties of Green Tea and originates in the West Lake Mountains in Hangzhou Jiangsu Province. The appearance is smooth, flat, straight, and shaped like a bird's tongue. The color is jade green and slightly yellow. The taste is sweet, fresh, and mellow, with a clean and high aroma. The tea soup is of green jade. The bottom of the leaves are tender.

碧螺春属卷曲形绿茶，产于江苏省吴县，其外形纤细卷曲呈螺状，嫩绿隐翠，清香幽雅，鲜爽生津。汤色碧绿清澈，叶底柔匀，饮后回甘，香气极为浓郁。

六安瓜片属于片形绿茶，产于皖西大别山区的六安市。成茶似瓜子状，故而得名。此茶色泽翠绿、香气清高、滋味鲜甘，十分耐泡，不仅可以消暑、解渴、生津，还有极强的助消化作用。

The appearance of Green Spiral Tea is curly and spiral. It is produced in Wu County, Jiangsu Province. It is tender green with a clean and elegant aroma. The taste is so brisk that it helps produce saliva and slake thirst. The tea soup is green and clear with soft and brewed leaves. It is a sweet aftertaste and mainly has a strong fragrance. Luan Guapian Tea is a Shape Green Tea from Luan of Dabie Mountain Region, Western Anhui. The tea leaves are in the shape of melon seeds. It is named. It is green jade with a clean and high aroma. The taste is fresh and sweet after several brewing times. It can not only relieve heat and quell thirst, but also avail the digestion rate and the curative effect.

绿茶的制作
Green Tea Processing Techniques

不发酵茶
Non-fermented Tea

绿茶是完全不发酵茶,与发酵茶和半发酵茶相比,叶绿素、维生素、茶多酚、咖啡因等天然物质的保留量较多。科学研究发现不发酵茶不仅可以抗过敏,还具有防止细胞老化、抑制癌细胞生长的功能,绿茶中含有的茶甘宁成分还能提高人体血管韧性,长期饮用有良好的保健作用。

Green Tea is non-fermented. Compared with fermented and partially fermented tea, its natural substance, such as chlorophyll, vitamins, tea polyphenols, and caffeine, is mainly retained. Scientific research has discovered that non-fermented tea can help anti-allergy, prevent cell aging, and inhibit cancer cell growth and proliferation. Green Tea also contains tea tannin, which could significantly improve the toughness of vascular. Therefore, it is suitable for health care if you drink for an extended period.

绿茶的制作与分类
Green Tea Processing and Classification

中国绿茶品种很多，造型又各具特色，不仅茶香耐人回味，还具有较高的欣赏价值。虽然各类绿茶造型工艺不同，但大致上都要经过杀青、揉捻、干燥三道工序流程。根据杀青和干燥方式的不同，可将绿茶划分为蒸青、炒青、烘青和晒青绿茶四种。

There are many kinds of Chinese Green Tea, and its shapes are different characteristics. Therefore, it not only has an aftertaste fragrance but also has high artistic value. Even the processing technique differs among Green Teas, and Green Tea generally has to go through three processing flows: fixation, rolling, and drying. Depending on the fixation and drying methods, the Green tea can be divided into four typs, such as steam fixation, stir fixation, hot air fixation, and sun fixation.

摊　青
Spreading Fresh Leaves to Dry

鲜叶采摘下来要先进行摊青。摊青需要在干净通风，相对湿度比较稳定的环境中进行。不同级别、不同品种的茶叶要分开摊放。因季节、气候、温度等不同，摊青的时间也略有不同，失水量达15%左右、茶芽变软即可。一般春茶需7～8个小时，夏茶要3～4个小时。

Those plucking fresh Green Tea leaves need to be dried first. This step

should be processed in a clean, dry, well-ventilated environment with more invariable air temperature and humidity. Different grades and varieties of tea leaves should be dried separately. Due to the variation of other seasons, climates, and temperatures. The time to driy is also slightry different, when the water loss reaches about fifteen percent and those tea buds soften. Spring tea typically takes seven to eight hours, and summer tea usually take three to four hours.

摊青过程中要适时翻晾叶片，以便散热。摊青可使嫩叶蒸发部分水分，有利于茶叶内含物的水解，降低茶的苦涩，并且为下一步的杀青减少了能耗和时间。

In the process of spreading green, we should timely turn over the tea leaves to cool them. Turning over can help to evaporate the moisture of tender leaves. In addition, it is beneficial to increase the hydrolysis degree of the inclusions to reduce bitterness of tea, and reduce energy consumption and time for the next step.

蒸　青
Steam Fixation

蒸汽杀青是中国古老的一种杀青方式，主要是利用蒸汽来降低酶的活性。从唐代就开始盛行，宋朝时得到进一步发展，后随着中华民族文化的传播传至日本，相沿至今。日本蒸青分玉露、煎茶、碾茶等，其中玉露属日本高级蒸青绿茶的特色。中国蒸青则以恩施玉露品质最为突出。

Steam fixation is an ancient method of sterilization in China. Steam is mainly used to reduce enzyme activity. It prevailed in the Tang Dynasty, rapidly developed in the Song Dynasty, and spread Chinese traditional culture to Japan. Craftsmanship has had a lasting impact throughout the ages. Those Japanese steam fixation teas, including Jade Dew Tea, Green Blade Sencha, and Flake Tea, are delicious. Jade dew is the charactenitics of Japanese advanced steamed green tea. Enshi Jade Dew and Cactus Tea are the best Chinese steam fixation tea types.

蒸青工艺保留了叶内较多的蛋白质、氨基酸、叶绿素、芳香物等物质，使绿茶形成茶叶深绿色、汤色嫩绿色、叶底青绿色的品质特征。通常蒸青绿茶的外形呈尖针状，茶香味醇而口感略显青涩。

The steaming fixation craft can retain tea's natural substances, such as protein, amino acids, chlorophyll, and aromatic. Making green tea formed in the quality characteristics of dark green tea, light green soup color, and green leaf bottom. Generally, the appearance of steaming fixation Green Tea is needle shape or platy shape. The aroma of this tea is fragrant with a pure taste. The taste is slightly astringent.

炒青
Stir Fixation

自明朝发明了制茶的炒青工艺之后，蒸青制法就逐渐被其取代了。炒青是以钢壁或滚筒壁的高温迅速破坏酶的活性，使茶多酚等停止氧化。为了避免红梗红叶现象的产生，温度一般达到180℃左右即可，然后再适当降低温度。叶内所含部分水分蒸发，增加了叶质的韧性和软度，为揉捻成形提供了条件，同时去除了茶的青草味，显露出茶的清香。

The steam fixation technique was gradually replaced by the stir fixation method, which was first invented in the Ming Dynasty. The high temperature of the steel bulkhead or drums wall could dramatically damage enzymes' activity to stop the oxidation of tea polyphenols. Avoid the appearance of the red stalk and red leaves, and the temperature should be controlled at about 180 degrees Celsius. Part of the water contained in the leaves evaporates, increasing the toughness and softness of the leaves, providing conditions of kneading and forming, while removing the grass taste of tea and revealing, the fragrance of tea.

烘　青
Hot Air Fixation

烘青根据不同制法可分为人工烘青和机械烘青。人工烘青多采用烘笼烘焙，先加热焙心，打毛火，焙心温度至90℃开始上茶，实行"高温薄摊，中间略厚"的原则。此过程要适时翻茶，叶子达五成干时，下焙摊凉；再打足火"低温慢烘"。机械烘青就是借助烘干机烘干，也有毛火、足火之分。烘青绿茶大部分用于窨制花茶的茶胚，干茶为墨绿色，清爽芬香。

Depending on the different processing methods, hot air fixation is divided into artificial hot air fixation and mechanical hot air fixation. The synthetic hot air fixation method works with the roaster baked. Heating the inner roaster first, then striking the first fire. Put those tea leaves into the roaster when the temperature rises to 90 Celsius. Thin flip should be processed at a high temperature and slightly thick in the middle. It would be best if you timely turned over tea leaves during this process. Bake the tea leaves below and cool when half the tea leaves are dry. Then strike the fire again to dry the tea slowly at a low temperature. The mechanical hot air fixation uses the fired dryer to dry tea leaves. It also has the first drying and final drying. Most Hot Air Fixation Green Tea is processed as the tea embryos to scent the Jasmine Tea. The dry tea is dark green with a tender flavor.

揉 捻
Rolling Tea Leaves

揉捻是绿茶塑形的一道工序，减小了茶叶的体积，绿茶的不同形态也是在此过程中显现的，为干燥成形奠定了基础。揉捻还能适当破坏部分茶叶细胞，使茶汁溢出附于叶表，茶味更加香醇。

Rolling is the process of Green Tea shaping. It could reduce tea size and create different Green Tea forms to lay the foundation for drying and forming. However, rolling can also damage partial leaf cells to spill over the juice to adhere to the leaf surface. As a result, the tea tastes more mellow and has a delicate fragrance.

揉捻分为冷揉和热揉两种。一般嫩叶容易成形，多采用冷揉，即杀青后摊凉，再进行揉捻，以此来保持叶的色泽。老叶纤维素含量高，宜采取热揉，即杀青后趁热揉捻，利于叶卷成条。

There are two ways of rolling, cool rolling and hot rolling. Cool rolling is used for those tender leaves, which are easily formed. After the green is spread cool, and then kneaded, so as to maintwn the color of the leaves. On the other hand, those old leaves with a high cellulose content need to be hot rolled. Rolling the tea leaves while hot would form the strip shape more quickly.

根据采取的方式不同，揉捻还可分为机械揉捻和手工揉捻。由于手工揉捻耗费人力，效率又低，所以除了龙井、碧螺春等手工名茶外，大多数茶叶都使用机械揉捻。

Rolling is generally divided into artificial and mechanical, according to the different methods. Manual rolling is labor-costing and inefficient, so most Green Teas are rolled by the machine, except for some famous handmade Green Longjing and Green Spiral Tea.

干 燥
Drying Tea Leaves

干燥是绿茶整形的工序，对经过揉捻的叶子整理、改进外形，蒸发掉多余的水分，使其便于运输和储存，并发挥茶香。绿茶的干燥有三种方式：炒干、晒干和烘干。一般分烘干和炒干两步进行。由于揉捻后的叶子还有一定的水分，所以需要先烘干，至水分蒸发到适宜锅炒时，然后再进行炒干。炒干更好地固定了茶的外形，提升了绿茶的清香。最后，经过炒干的绿茶起锅摊凉即完成全部制作。

Drying is the procedure of sorting out and shaping Green Tea. Improving tea leaves' appearance by evaporating excess moisture in leaves is for easy packing, shipment, storage, and volatilizing the aroma. There are three drying means for Green Tea, stir fixation, sun fixation, and hot air fixation. It is usually with two steps: hot air fixation and stirs fixation. After rolling, tea leaves still have some water, and it needs to be dried first to evaporate excess moisture, then stir-fried. Stir fixation can better fix the shape and enhance the aroma of tea leaves. The last step is cooling to finish the whole process.

白茶

White Tea

白茶是中国茶叶中的特殊珍品，一般地区并不常见。白茶在中国历史悠久，北宋时期便有种植。茶毫颜色如银似雪，汤色黄绿清澈，香气清鲜。滋味清淡回甘，令人回味无穷。

White Tea is a unique treasure of Chinese tea. This rare plant was cultivated during the period of the Northern Song Dynasty. The color of the fuzz is as white as silver and snow. The tea soup is yellowish-green and clear with a fresh fragrance. The tea taste is plain and unforgettable after taste.

白茶最显著的特点是富含氨基酸，特别是高含量的茶氨酸，不但能提高成品茶的香气和鲜爽度，还能提高人体机能的免疫力，有利于身体健康。尤其是陈年的白毫，有着防癌、抗癌、防暑、解毒、缓解牙痛的功效。

The most noteworthy feature of White Tea is rich in the amino acid tryptophan, especially Theanine's high content. Therefore, it can enhance the fragrance, increase the freshness and briskness of commercial tea, and improve the immunity of human function. Moreover, it is suitable for health. The crusted white fuzz tea has the effects of cancer prevention, anticancer, heatstroke prevention, detoxification, and toothache treatment.

白茶的品质
Quality of White Tea

芽茶和叶茶
Tip-Tea Type and Leaf-Tea Type

白茶因茶树品种、原料鲜叶采摘的标准不同，分为芽茶和叶茶。白芽茶的典型代表当属白毫银针，产地主要集中在福建福鼎、政和两地。白芽茶具有外形芽毫完整，满身披毫，香气清鲜，汤色黄绿清澈，滋味清淡回甘等品质特点，属轻微发酵茶，是中国茶类中的特殊珍品。

The tea plant variety and a different standard of fresh raw tea plucking white tea is divided into the tip-tea and leaf-tea types. White Tip Silver Needle is the typical representative tea of white tip tea. The origin of this tea is mainly in Funding City and Zhenghe City, Fujian Province. White bud tea has complete buds and is covered by the hairy tip. The scent is clear and fresh. The tea soup is yellowish-green and clear. And the white tip tea.

White bud tea is light fermentation tea. It has complete buds and is covered by the hairy tip, the scent is clear and fresh, the tea soup is yellowish-green and clear, and the white bud tea with a light and sweet taste. It belongs to the special trevsures of Chinese tea with a light taste and sweet aftertaste. It is the unique treasure of Chinese tea.

叶茶的代表有白牡丹、新工艺白茶、贡眉、寿眉等，成品茶带有特殊的花蕾香气。

The representative teas of white-leaf tea are White Peony Tea, New Process White Tea, Tribute Eyebrow Tea, and Long Life Brow Tea. The finished tea product has a special flower bud aroma.

名茶种类
The Variety of Chinese Famous White Tea

白毫银针属于白芽茶，是白茶中的极品。用一芽一叶的肥壮芽头制成，成茶遍披白毫，挺直如针，色白如银，香气清新，滋味甜爽，汤色呈浅杏黄色。

White Tip Silver Needle Tea is the best grade tea among the White Tip-Tea, made from those bold buds of tea leaves. The finished tea product is covered with white fuzz. It is as straight as a needle, and the color is as white as silver. The tea aroma is fresh and clean, sweet and has brisk taste. The color of the tea soup is light apricot yellow.

白牡丹属白叶茶，因其干茶呈绿叶夹银毫状，冲泡后绿叶夹着嫩芽，宛如牡丹初绽而得名。贡眉也属白叶茶，优质贡眉芽显毫多，色泽绿，汤色橙黄或深黄，香气馥郁，滋味甘爽，叶底灰绿明亮。

White Peony Tea is a kind of white-leaf tea. Those tea leaves are green, with plenty of silver fuzz. Brew tea leaves with tender buds look like the early bloom of a peony, hence its name. Tribute Eyebrow Tea is a kind of white-leaf tea too. The higher-grade Tribute Eyebrow Tea has plenty of white fuzz. The color is green. The tea soup is orange-yellow or deep yellow. It has a strong fragrance and a sweet and brisk taste. The brew tea leaves are greyish-green and bright.

品质鉴别 Quality Identification

白茶属于微发酵茶，是中国六大茶类的一种。因茶树品种和产地的严格限制，白茶的品种少、产量低，因此优质白茶显得尤为珍贵，品质的优劣也较容易辨别。根据制作原料的不同，白茶主要分为五个品种：白毫银针、白牡丹、贡眉、寿眉和新工艺白茶。白毫银针由肥壮毫芽制成，不带梗蒂，品质最佳；其余四种由细嫩芽叶制成，以芽多而肥壮者为上品。

White Tea is a kind of light fermentation tea and one of China's six major tea categories. There are few types and low White Tea yields because of those strict requirements on tea varieties and growing regions. Therefore, high-quality White Tea is precious. And the quality grade is easier to distinguish. According to the different processing materials, White Tea has five varieties. They are white silver needle, white peony, Tribute Eyebrow Tea, Long Life Brow Tea, and New Process White Tea. White Tip Silver Needle Tea is made from bold buds without the stalk. The quality is the best. The rest four types are made from tender bud leaves. The high-grade tea is with plenty and fat buds.

白茶的审评侧重于外形，不同品种在白茶品质特性的表现上均以芽多毫显、叶张肥嫩、色泽灰绿或褐绿者为上品；芽稀而瘦小或无芽，叶张单薄，色泽棕褐发灰的为下品。其中，白牡丹以叶态伸展的为好，新工艺白茶则以条索粗松带卷的更佳。

The evaluation method of White Tea is focused on the tea's appearance. The quality characteristics of different varieties of White Tea are full of tea buds with white fuzz. Tea leaf is fat and tender. The greyish-green and brownish-green color tea have higher grades of White Tea. Tea leaves are with few buds or without buds. Brownish auburn and greyish thin tea leaf are low-grade White Tea. The higher-grade white peony has a flat leaf side. The higher grade of New Process White Tea has coarse, loose, and curly tea leaves.

从白茶冲泡后的内质来看，以毫香浓郁、汤色浅黄、滋味醇厚鲜爽、叶底嫩匀齐整的为好；香气单薄、汤色暗浊、滋味粗涩、叶底粗老、叶张破损的为差。

The fragrance of fuzz is strong and heavy of the brewed White Tea. The tea soup is light yellow with a mellow and brisk taste. The brew leaves are tender and even of the higher grade White Tea. The lower grade White Tea has a weak fragrance, lots of suspensions, course brew tea leaves, and broken tea leaves.

白茶的制作
White Tea Processing Techniques

萎凋的作用
The Effect of Withering

　　萎凋是白茶制作中最关键的工序，是形成白茶银白光润的色泽，清新淡雅的茶香，甘醇鲜爽的口感这三大独特品质的重要环节。白茶萎凋不仅能蒸发掉鲜叶内的水分，还能使叶内的物质发生化学变化，水分蒸发先快后慢，直到完全干燥。萎凋前期，酶的活性增强，叶内有机物水解，多酚类物质发生氧化。随着萎凋的进行，水分减少到一定程度时，酶的活性逐渐下降，氧化受到抑制，从而有效地去除了茶的苦涩和青气。

　　The step of withering is an essential procedure of White Tea processing techniques. It is also necessary to form three unique characteristics: silvery color, fresh and gentle tea aroma, and mellow, brisk taste. The withering of White Tea not only evaporates the water in fresh tea leaves but also makes

the chemical change of leaf substance. The moisture evaporation is faster at the first stage and slower at the second stage until complete drying. The enzymatic activity of the enzyme is enhanced in the first phase of withering. The organic matter is hydrolyzed inside the tea leaf. Its polyphenols are also oxidized. With the withering going on, the enzyme activity decreases continuously when the moisture is reduced to a certain extent. The oxidization will be inhibited to remove the tea's bitterness and green odor.

萎凋的三种方式
Three Methods of Withering

白茶萎凋有室内自然萎凋、复式萎凋和加温萎凋三种。室内自然萎凋是将鲜叶摊放放在筛内摇匀，静置35～45小时，至七八成干，叶芽毫色发白，叶色变为深绿，稍有卷翘即可，这一步骤俗称开青。以此种萎凋制成的白茶品质最佳。复式萎凋多适用于春茶，在阳光温和的天气里，将开青后的鲜叶放在较弱的日光下晒10～20分钟，待叶子失去光泽，再转移到室内萎凋。加温萎凋解决了多雨季节不宜萎凋的困难，温度应控制在30℃左右，相对湿度65%～75%为宜。

There are three methods: natural withering, natural withering combined with sun-fixed withering, and indoor natural withering and warming withering. Put the fresh tea leaves into the sifter and shake them up. Then set them for 35 to 45 hours until 70% and 80% dryness level of leaves until the bud tea leaves are covered with white fuzz and their color turns dark green. Finished processing until the tea leaf is slightly curly. This step is commonly known as stirring prepared. The quality of this White Tea is the best with this method of withering. The process of Natural Withering combined with sun-fixed withering is suitable for Spring Tea. Put the fresh leaves after withering to dry in the sun for ten to twenty minutes in the gentle weather. Move tea leaves inside for indoor withering when the leaves are tarnished. The warming withering can resolve the difficulties in the rainy season. It is proper to control the temperature within about 30℃ and 65% ~ 75% relative humidity.

干　燥
Drying

萎凋完成后的白茶应立即进行干燥。干燥对白茶有定色和提香的作用，并能充分去除水分，防止茶叶变色变质。干燥的方式有烘笼烘焙和干燥机烘焙。烘笼烘焙在萎凋程度一般达八九成干时进行，温度应掌握在90℃左右，烘10 ~ 20分钟。若萎凋程度只有六七成干，则需进行二次烘焙。烘焙时需注意翻叶要轻，避免叶芽碎断，降低成茶品质，烘焙温度为80℃ ~ 90℃。

Drying should be done at once after withering White Tea. The drying step can help instant set color and enhance the aroma of White Tea. It also

can entirely remove water to prevent changing color and deterioration. The drying methods can be divided into bamboo-frame drying or machine drying. First, it can be processed until the withering degree is up to 80 percent to 90 percent. And the temperature should be controlled at around 90℃. Next, it should be baked for 10 to 20 minutes. It needs to be baked again if the withering degree is only 60 percent to 70 percent. Then gently turn over the leaves to avoid damage to leave buds and debase the tea quality. Again, the temperature should be controlled between 80℃ and 90℃.

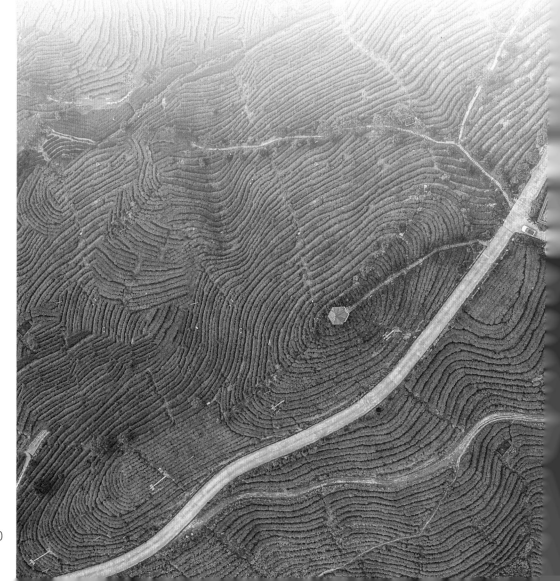

干燥机烘焙是将萎凋七八成干的叶子分两次进行烘焙，摊叶厚度 4 厘米。初焙速度要快，温度 100℃ ~ 110℃，约 10 分钟。复焙调慢速度，温度为 80℃ ~ 90℃，约 20 分钟即可焙至足干。

Tea leaves with 70 percent to 80 percent dryness should be baked with the machine dryer in two steps. First, the thickness of the spreading leaves should be four centimeters. The baking time should be shorter, and the temperature should be 100℃ ~ 110℃ for about 10 minutes. The second-time baking can be processed for 80℃ ~ 90℃ slowly. Then let it dry for twenty minutes.

黄茶
Yellow Tea

　　黄茶色泽金黄光亮，最显著的特点就是"黄汤黄叶"。茶青嫩香清锐，茶汤杏黄明净，口味甘醇鲜爽，入口回甘，收敛性弱。以君山银针为代表的黄茶在国内和国际市场上都久负盛名，现在已是身价千金。黄茶性凉微寒，所以适合胃热者饮用。夏季天气酷热，选择黄茶也可起到适当的祛暑解热之功效。

　　Yellow Tea is golden bright, with the most distinguishing feature of the natural light yellow color of tea soup and tea leaves. Those fresh leaves are tender with a fresh and sharp fragrance. The tea soup is apricot yellow, clear, and bright, with a sweet and mellow taste after the taste. The Green Tea astringency is weak. Jun Mountain Silver Needle Tea represents Yellow Tea in the domestic and overseas markets. It has curiosity value today. The character is slightly cold of Yellow Tea in Chinese traditional medicine. So drinking is

suitable for people, especially those with stomach-heat syndrome. The summer is sweltering. So drinking Yellow Tea also can reduce summer heat and relieve fever during the season.

黄茶的品质
Quality of Yellow Tea

黄芽茶
Yellow Bud Tea

"闷黄"工序是黄茶独有的加工方法,可使黄茶具有黄汤黄叶的特色。黄茶的分类标准是按照鲜叶的嫩度和芽叶大小。黄芽茶原料细嫩,是采摘最细嫩的单芽或一芽一叶加工制成,幼芽色黄而多白毫,故名黄芽,滋味鲜醇。

The sweltering step is a unique processing method of yellow tea which makes yellow tea have the characteristic of yellow soup and yellow. The classification standard of Yellow Tea is according to the fresh leaf tenderness degree and leaf shape size. Yellow bud tea leaves are delicate and tender. It is made from the most delicate single buds or a bud with a leaf. The leaf bud is yellow with plenty of fine silvery hair. It is the name of Yellow Bud Tea.

黄茶由于品种的不同，在叶片的选择和加工工艺上有相当大的区别，最有名的品种包括湖南岳阳洞庭湖的君山银针、四川雅安的蒙顶黄芽和安徽霍山的霍山黄芽。

Simultaneously, there is a big difference in the processing technology and selection of material tea with a different variety. The most famous variety list includes Jun Mountain Silver Needle from Dongting Lake Yue Yang City in Hunan Province, Mengding Yellow-bud Tea from Yaan City in Sichuan Province, and Huoshan Yellow Bud Tea from Huoshan in Anhui Province.

黄小茶
Yellow Little Tea

黄小茶是采摘细嫩芽叶加工而成的，一芽一叶，条索细小。目前国内产量不大。主要品种有湖南岳阳的北港毛尖，湖南宁乡的沩山毛尖，湖北远安的远安鹿苑和浙江温州、平阳的平阳黄汤。

The Yellow Little Tea is processed with those tender buds. The leaf is small with a tea bud. Domestic production today is not significant. The main varieties of Yellow Little Tea are Beigang Tip from Yueyang city in Hunan Province, Weishan Tip from Ningxiang city in Hunan Province, Luyuan Tea in Yunnan County Hubei Province, and Pingyang Yellow Tea in Pingyang County Wenzhou City in Zhejiang Province.

沩山毛尖芽叶肥硕多毫，色泽黄亮油润，白毫显露，汤色橙黄明亮，滋味甘醇爽口，叶底黄绿嫩匀，带有一种特殊的松烟香。远安鹿苑呈条索环状，色泽金黄，略有鱼子泡，冲泡后香郁高长，滋味醇厚，入口回甘，汤色黄净明亮。

The bud leaf of Weishan Tip is fat with plenty of fuzzes. The color is yellow bright, and bloom. Lots of white fuzz appear with it. The tea soup is orange-yellow bright with a sweet and mellow taste. The brewed leaves are yellowish-green, tender, and have unique pine smoky flavors. The tea shape of Luyuan Tea is cyclic annular. Its color is golden yellow with some scorched points. The high aroma can last after brewing with a mellow and thick taste. The sound-quality tea is a sweet aftertaste. The tea soup is yellow, clear, and bright.

黄大茶
Large-Leaf Yellow Tea

黄大茶是中国黄茶中产量最多的一类，主要产于安徽霍山及邻近的湖北英山等地，距今已有400多年历史，其中以安徽的霍山黄大茶、广东的"大叶青"品质为佳，最为著名。

Large-Leaf Yellow Tea is the most significant production variety among Chinese Yellow Teas. And it is mainly grown in Huoshan City in Anhui Province and its neighboring area Yingshan in Hubei Province. It has a history of more than four hundred years since then. The quality of Huoshan Yellow Tea and Dayeqing Tea from Guangdong Province are the best among these varieties.

黄大茶的鲜叶采摘要求大枝大杆，一芽四五叶，长度在 10 ~ 13 厘米。春茶一般在立夏前后开采，为期一个月。夏茶在芒种后开采，不采秋茶。制法分杀青、揉捻、初焙、堆积、拉小火和拉老火等几道工序。黄大茶的特点是叶大梗长、叶片成条，梗叶相连，形似鱼钩，梗叶金黄油润，汤色深黄偏褐色，叶底也是黄中显褐。黄大茶味浓厚，耐冲泡，具有突出高爽的焦糖香味。

Therefore, it is required to pick up tea leaves with large branches for processing Large-Leaf Yellow Tea. One tea bud should have four or five leaves the length 10 to 13 centimeters long. Spring tea is typically picked up during the solar ferms at the Beginning of Summer. It will last for the whole month. Summer tea is picked up after the time of grain in the ear. Do not pick up tea in autumn. Processing methods can be divided into fixation, rolling, first-time baking, piling up, drying with a small fire, baking with a low fire, and then baking with a high fire in a short time. The characteristics are as follows: big size leaf with a long stem, leaf with yellowish stripes, and stalks that look like hooks. Those golden fresh tea leaves can be brewed several times. The brewed leaves are yellow with dark brown and taste rich and mellow. It has a high and lightly cool Caramel fragrance.

名茶种类
The Variety of Chinese Famous Yellow Tea

黄芽茶中的极品当属湖南洞庭湖出产的君山银针，其外形茁壮挺直，重实匀齐，银毫披露，金黄光亮，内质毫香鲜嫩，汤色杏黄明净，滋味甘醇鲜爽，在国内外都久负盛名，身价极高。

The best grade of the Yellow-bud Tea is the Jun Mountain Silver Needle of Dongting Lake in Hunan Province. The fresh tea leaf is the study and straight. Its heavy body is growing evenly. This tea leaf is fat and straight with silver hair. It is golden, bright, fresh, and tender. The tea soup is apricot yellow, clear, and bright. The taste is sweet, mellow, and brisk. It enjoys high fame in the world with its high status.

另一代表花色的是蒙顶黄芽，其外形偏直，肥壮匀齐，色泽金黄，汤色黄亮，甜香浓郁，滋味浓醇，叶底嫩黄匀亮。

Another suitable representative type is Mengding Mountain Yellow Bud. The dry tea leaf is straight, fat, and bold. The color of tender and even leaves are golden. The tea soup is yellow and bright, with a strong sweet aroma and mellow taste. The brewed tea is tender, even, and bright.

安徽霍山黄芽亦属黄芽珍品，其形如雀舌，芽叶细嫩多毫，色泽嫩黄，茶汤有板栗香，饮之口有回甘。霍山大化坪金鸡山的金刚台所产的黄芽最为名贵，干茶色泽自然，呈金黄色，香高、味浓、耐泡。

Huoshan Mountain Yellow Bud is also a treasure type of Yellow Tea. Its dry tea leaf is shaped like a bird's tongue. The bud tea leaf is tender with plenty of fuzzes and delicate yellow color. Tea soup has a chestnut flavor. It is a sweet aftertaste. The most precious yellow bud tea from Jingangtai in the Huoshan Mountain area is yellow bud tea. The color of dry tea is golden, with a high aroma and strong flavor. It can be brewed many times.

北港毛尖是黄小茶中的名茶，其外形芽状叶肥，毫尖显露，呈金黄色，汤色橙黄，香气清高，滋味醇厚。

Beigang Maojian Tea is a famous type of small-leaf Yellow Tea. Its golden bud leaf is fat with tippy and orange-Yellow Tea soup. The fragrance is clear and high, with a mellow and thick taste.

黄茶的制作
Yellow Tea Processing Techniques

杀 青
Fixation

　　黄茶属于后轻发酵茶，它的杀青与绿茶原理基本相同，但温度比绿茶稍低一些，且时间较长。杀青时要多闷少抛，创造高温湿热的环境，以破坏叶细胞中酶的活性，使叶绿素受到较多损害，多酚类物质发生自动氧化和异构化。

　　随着叶内所含的淀粉、蛋白质分解为单糖和氨基酸，部分水分蒸发，杀青在提高了茶芳香的同时也发散了青草味和苦涩的口感。这是形成黄茶"黄汤黄叶"特点的前提条件。

　　The fixation step for Yellow Tea (light post-fermented tea) and Green Tea is basically on the same principle. But the fixation temperature for Yellow Tea is slightly lower, and the time is longer than for Green Tea. It needs to be sweltered for more time but with less stirring. Create a high temperature and humid environment to destroy the enzymes' activities in the leaf cell. So that the chlorophyll is more damaged, and the polyphenols will auto-oxidation. Isomerization will occur for the polyphenols at the same time.

As the starch and protein inside the leaves are broken down into monosaccharides and amino acids by evaporating partial water, the fixation step enhances the tea fragrance and creates a bitter taste. Therefore, it is a prerequisite for making the characteristics of Yellow Tea soup and leaves for Yellow Tea.

闷黄是黄茶制作特有的一道工序。根据黄茶种类的差异，进行闷黄的时间先后也不同，可分为湿坯闷黄和干坯闷黄。如沩山毛尖是在杀青后趁热闷黄，温州黄芽是在揉捻后闷黄，属于湿坯闷黄，水分含量多且变黄块。打晃茶则是在初干后堆积闷黄；君山银针在炒干过程中交替进行闷黄；霍山黄芽是炒干和摊放相结合的闷黄，称为干坯闷黄，含水量少，变化时间长。叶子含水量的多少和叶表温度是影响闷黄的主要因素。湿度和温度越高，变黄的速度越快。闷黄是形成黄茶金黄的色泽和醇厚茶香的关键工序。

The procedure of sweltering is an unusual process step of the Yellow Tea. According to the varieties of Yellow Tea, the dampness in the order is also different. It can be divided into wet body sweltering and dried body sweltering, such as the Weishan Tip sweltered after fixation while it was warm. Wenzhou Yellow Tea is sweltered after rolling. It is called wet leaf body sweltering. The moisture content is high, and it can turn yellow quickly. Large-leaf Yellow Tea is sweltered with a stack-cover process after the first drying. The sweltering step of the Jun Mountain Silver Needle alternates in the process of stir fixation. The stir fixation and tedding step are combined with the sweltering step for Huoshan Yellow Bud tea. It is called dried leaf body sweltering. The moisture content is low, and the leaf could take too long to convert. The moisture content and leaf temperature are the main factors affecting sweltering. It turns yellow faster with higher humidity and temperature. The sweltering step is the critical process of creating the golden yellow color and mellow taste of Yellow Tea.

干 燥
Drying

黄茶的干燥比其他茶种温度要低，一般采用分次干燥，即毛火烘干和足火炒干。毛火温度较低，水分蒸发缓慢，干燥的时间相对较长，有利于叶组织内含物的转化，多酚类物质的自动氧化，进一步增强了叶子的黄变，巩固了茶的色泽。足火温度略高，促进了单糖和蛋白质的转化，高沸点芳香物质的发挥，增进茶香。温度先低后高是形成黄茶独特香味的重要因素。

The drying temperature for Yellow Tea is lower than other kinds of tea. It is generally dried by stages, that is, dry by hair fire and dry by fool fire. The temperature is low during the first drying, and the moisture evaporation is slow. Drying for a relatively long time benefits transforming inclusion in leaf tissue. The auto-oxidation of polyphenols can further enhance the leaf's yellowing to maintain the tea's color. The temperature is slightly higher for the final drying. It promotes the conversion of monosaccharides and proteins. The volatilization of the high-boiling point aromatics can enhance the aroma of tea. The temperature from low to high is essential in creating a unique fragrance of Yellow Tea.

乌龙茶的特点介于红茶与绿茶之间,其综合了红茶和绿茶的制作方法,既保持有红茶的浓鲜味,又有绿茶的清芬香。茶叶在水中呈"绿叶红边",品尝后齿颊留香,回味甘鲜。

The character of Oolong Tea is between Black Tea and Green Tea. It summarizes the different processing methods of Black Tea and Green Tea. Oolong Tea can keep the rich and fresh taste of Black Tea and has the delicate fragrance of Green Tea. Green Tea leaves have red edges in the water. Oolong Tea has a delicious and distinctive aroma, which is hard to resist. It displays freshness and sweetness in the aftertaste.

乌龙茶的品质
Quality of Oolong Tea

闽北乌龙茶
The North Fujian Oolong Tea

闽北乌龙茶主要是岩茶，产于福建武夷山一带，主要有武夷岩茶和闽北水仙，其中又以武夷岩茶最为著名。因武夷山的生态环境极为适宜茶树生长，再加上其独特精湛的制作工艺，使得武夷岩茶驰名中外。武夷岩茶外形匀整，壮结卷曲，色泽青翠润亮，叶背呈蛙皮状沙粒白点，冲泡后汤色较深，叶底、叶缘显朱红，中央呈浅绿色，红绿映衬，形成奇特的"绿叶红镶边"。品饮此茶，香气馥郁，滋味浓醇，鲜滑回甘，"锐则浓长，清则幽远"，具有特殊的"岩韵"。代表名茶有大红袍、肉桂、铁罗汉等。闽北另一花色水仙茶的品质也别具一格，有"茶质美而味厚，奇香为诸茶冠"的美誉。

The North Fujian Oolong Teas are mainly Rock tea. They are primarily made in the Wuyi Mountain area in Fujian Province. Two major types are Wuyi Rock Tea and North Fujian Shui Xian Tea. Wuyi Rock Tea is the most famous type. With the Wuyi Mountain area's excellent ecological environment and the unique and exquisite processing craft, Wuyi rock tea is renowned worldwide. Those dry tea leaves are neat and evenly. The appearance is bold and curly. The color is green, jade, and blooming. There are some frog skin lines with powdery white spots underneath tea leaves. The tea soup color is deep, and the edge and body of the brewed tea leaves are bright red after brewing. It is light green in the middle of the leaves. The red edge and green leaves set each other off beautifully. It produces a particular effect of green leaves with red edges. The tea has a fragrance with a mellow and heavy taste. It is a sweet and smooth aftertaste. The sharp aroma is rich and lasts for a long time. The fresh aroma is gentle and far away. This tea has a particular YEN flavor. The famous representative teas are Da Hong Pao, Rou Gui, Tie Luo Han, etc. Another Shui Hsien Oolong Tea character in North Fujian is also unique. It also won the high praise and acclaim of the majority of drinkers. Tea is good with a mellow and heavy taste. The delicate flavor is the best among all kinds of tea.

闽南乌龙茶
The South Fujian Oolong Tea

闽南乌龙茶主要是铁观音,源于闽北武夷山,后在闽北乌龙茶的基础上汲取长处不断发展,形成了自己独有的制作工艺。其制作严谨,技艺精巧,对茶树鲜叶采摘的成色、采摘时间、天气、制法,都有极为精确的要求,在国内外茶叶市场上享有盛誉。

The most representative South Fujian Oolong Teas are Tie Guan Yin. It originates from the Wu Yi Mountain area in North Fujian. The unique processing craft was created and developed based on North Fujian Oolong Tea. It has conscientious production and a unique style. The South Fujian Oolong Tea only can be accomplished with skill and ingenuity. It has been shown that the accuracy can meet the color leaves standard, plucking time, and processing method. South Fujian Oolong Tea has a worldwide reputation in the tea market.

广东乌龙茶
Guangdong Province Oolong Tea

广东乌龙茶主要产于广东潮汕地区，加工方法源于福建武夷山，因此其风格流派与武夷岩茶有些相似。凤凰单枞和凤凰水仙是广东乌龙茶中的优秀产品，历史悠久，品质特佳，为外销乌龙茶之极品，闻名中外。它具有天然的花香，卷曲紧结而肥壮的条索，色泽青褐而带红线，汤色黄艳带绿，滋味鲜爽而浓郁甘醇，叶底绿叶红镶边，耐冲泡，连冲十余次，香气仍然溢于杯外，甘味久存，真味不减。近年来，广东的石古坪乌龙茶和岭头单枞也迅速崛起，为广大茶人所喜爱。

Guangdong Province Oolong Tea mainly grows in the Chaozhou-Shantou Region of Guangdong Province. The processing method is unique and originates from the Wuyi Mountains area in Fujian province. Therefore, its style and spirit have some similarities with Wuyi rock tea. Fenghuang Dan Cong and Fenghuang Shui Xian tea are the best products of Guangdong Oolong Tea. Tea production has a long history of excellent quality. It is the highest-grade product of exportable Oolong Tea and is well-known inside and outside the country. It has a natural flowery flavor. The curly and tight leaf is also bold. Tea is blueish auburn with a red edge. And tea soup is yellow bright, and green, the taste is fresh, rich and mellow the brewed leaves are green with a red edge, brew resistance even more than ten times, the aroma is still overflowed in the cup, sweet lasting, the true taste is not reduced. In recent years, Shiguping Oolong Tea and lingtou Danzoing Tea in Guangdong have also risen rapidly, and are favored by the majority of tea people.

台湾乌龙茶
Taiwan Province Oolong Tea

台湾乌龙茶产于台湾地区，是自福建安溪移植而来，依据其发酵程度不同，可分为轻发酵乌龙茶、中发酵乌龙茶及部分重发酵乌龙茶三类。清香乌龙茶及部分轻发酵包种茶属轻发酵乌龙茶，其品质特征是色泽青翠，冲泡后汤色黄绿，花香显著，叶底青绿，基本上看不出有红边现象。

Taiwan Province Oolong Tea is produced on Taiwan Island and transplanted from Anxi County of Fujian Province. The different fermentation degrees can be divided into three categories: light fermentation Oolong Tea, medium fermentation Oolong Tea, and partial heavy fermentation Oolong Tea. Delicate fragrance type Oolong Tea and some light Oolong (Pouchong) belong to light fermentation Oolong Tea. Its quality characteristics include green jade dry tea and yellowish-green Tea soup with a strong floral fragrance. The brewed leaves are blueish-green. The red edges of tea leaves can not be seen clearly.

中度发酵乌龙茶主要有冻顶乌龙、木栅铁观音和竹山金萱等。外形多数为半球状颗粒，也有卷曲状。其色泽青褐，汤色金黄，有花香和甜香，滋味浓醇，叶底多数黄绿，可见少量红边。

重度发酵的乌龙茶有白毫乌龙，色泽乌褐，嫩芽有白毫，汤色橙红，有蜜糖香和果味香。

Medium fermentation Oolong Tea mainly includes Dongding Oolong Tea, Mushan Tie Guan Yin, Zhushan Jinxuan, etc. It is shaped like a hemisphere. There are also some curly tea leaves. The dry tea is blueish auburn. The tea soup is golden yellow with a flowery and sweet flavor. The taste is heavy and mellow. Most brewed tea leaves are yellowish-green. A few red edges could be seen.

Whitetip Oolong is one kind of heavy fermentation Oolong Tea. The color is auburnish black. The tender bud has silvery downy feathers. The tea soup is orange-red with a honey-sugar and fruit flavor.

香型类别
Fragrant Type

乌龙茶有"中国特种茶"之称，其花色品种丰富，名优茶种类众多，从香型上主要分为两类：清香型乌龙茶和浓香型乌龙茶。

Oolong Tea is a kind of unique Chinese tea. There are variations in colors and styles. It is divided into two categories: delicate fragrance type Oolong Tea and strong aromatic Oolong Tea.

"清香型"乌龙茶又名"台式"乌龙茶，主要是台湾地区在安溪乌龙茶的基础上，以独特的栽培和加工制作技术生产出来的自成一脉的乌龙茶。清香型乌龙茶表现出来的特质有：干茶呈球形或半球形，色泽碧绿，冲泡后在杯中呈茶蕾状，香气清新持久，汤色明亮黄绿，口感鲜嫩回甘，韵味强，叶底浓绿柔软。其代表品种冻顶乌龙是乌龙茶中的后起之秀。

The delicate fragrance type of Oolong Tea is also known as Taiwan-type Oolong Tea. It is sui generis and developed with unique cultivation and processing techniques based on the Anxi Oolong Tea. This type of Oolong Tea has some peculiar traits. The appearance of dry tea is usually spherical or hemispherical and green jade. The brewed tea is the shape of a bud with a long-lasting fresh fragrance in the cup. The tea soup is bright yellowish-green. The taste is fresh and mellow with a strong flavor. The brewed tea is dark green and soft. The representative variety is Dongding Oolong Tea. It represents a new rising star among the Oolong Tea.

"浓香型"乌龙茶以传统工艺生产制作，相对于清香型乌龙茶有做青程度较重、烘焙时间较长等细微区别，不同种类的名优茶品还有各自独特的工序和工艺要求。其主要特质有：条索粗壮紧结，

重实匀整，色泽绿润，有光泽，香气浓郁，深沉持久，滋味醇浓清爽，回味悠长，汤色橙黄艳丽，叶底黄绿镶红边，有"七泡有余香"之称。代表品种武夷岩茶，具有"岩骨花香"的独特韵味。

The strong aromatic type of Oolong Tea is produced with the traditional technique. There are some subtle distinctions between these two kinds. Compared with the delicate fragrance type Oolong Tea, the rolling degree is higher, and the baking time is longer. Different kinds of famous tea have their unique processing and technological requirements. It shows its main characteristics, such as a bold leaf, heavy body, and evenly divided. The color is green, blooming with a higher luster degree. It also has a heavy and mellow flavor and can last for a more extended period. The taste is the mellow and rich, long aftertaste. The tea soup is bright orange-yellow. It produces an effect of yellowish-green leaves with red edges. People can enjoy lingering fragrances after seven brewing times. The representative variety is Wuyi Rock tea. It has the unique charm of the YEN flavor.

乌龙茶的制作
Oolong Tea Processing Techniques

部分发酵茶
Semi-fermented Tea

部分发酵茶也称乌龙茶，又称青茶。其特性介于全发酵茶和不发酵茶之间，既具有红茶的醇香，但无热性；又具有绿茶的清爽，却不青涩。因叶子中心显绿，叶边发红，故有"绿叶红镶边"的美称。乌龙茶内含有丰富的茶多酚，能有效分解脂肪，达到减肥的功效。乌龙茶的主要产地有福建、广东、台湾等。制作工序大致可分为：萎凋、做青、杀青、揉捻和干燥五个步骤。

Semi-fermented tea is also called Oolong Tea and bluish-Green Tea. Its features and characteristics are between fully fermented tea and non-fermented tea. This kind of tea is pure and mellow Black Tea and has the crisp flavors of Green Tea. Its nature is not hot and without astringent taste.

It has a reputation for green leaves, with a red edge for its green center leaf and red leaf edge. Therefore, it has the reputation of "green leaves with red edge". Oolong Tea is full of tea polyphenols to break down fats efficiently. It could help you lose weight. The main producing areas are Fujian, Guangdong, Taiwan, etc. The main processing procedure can be classified into five steps: withering, fermentation, fixation, rolling and drying.

室外萎凋
Outdoor Withering

乌龙茶的萎凋与发酵几乎是同时进行的。萎凋分为室外萎凋和室内萎凋。室外萎凋又称日光萎凋，也就是通常所说的晒青。鲜叶采摘下来后，为防止茶青闷坏，应立即摊晒散热，根据气温的高低适时翻晒。一般来说晒青应在较弱的阳光下进行，夏季的午后光线太过强烈，不宜摊晒，以免芽青因温度过高被灼伤。待茶青变软后，即可移至室内。

The withering and fermentation of Oolong Tea are almost going on simultaneously. The withering can be divided into outdoor withering and indoor withering. Outdoor withering is also called sunlight withering, known as sun fixation. They avoid covering tightly with tea leaves. Instead, they should be spread out to cool. Turn it in time according to the temperature. In general, sun fixation should be processed under the sun. It should not be flatted to dry in the summer afternoon. The sun is too strong. To avoid green buds would be burned at too high a temperature, leaves can be moved to an indoor place when green buds soften.

室内萎凋
Indoor Withering

室内萎凋也称凉青。茶叶移至室内后，需要静置一段时间，使水分均匀分布，同时适当翻动，促进水分蒸发，再静置，再翻动……如此循环数次，直至达到理想干度为止。这就是俗称为"走水"的过程。

Indoor Withering is also called cooling. After the tea is moved indoors. Then cool down for a period to make the moisture distribution even more. Keep stirring tea leaves properly to promote fast evaporation. Then cool and stir, relapse several times until the desired dryness is reached. It is usually called the process of the moisture removal step. This is the process known colloqwially as "water walking".

萎凋时茶内水分逐渐散失，叶细胞膜的半透性遭到破坏，酶的活性增加，叶内含物开始转化，去除了茶中的苦涩及青气味，使香气逐渐显露出来，并促进茶叶的发酵。

The moisture inside the tea is gradually evaporated while withering. The semi-permeability of the cell membrane is damaged. The activity of enzymes is increased. It also helps promote the further conversion of matters contained in the tea. It removes the bitter taste and astringency of tea and shows the aroma. It also can help to promote fermentation.

做青和摇青
Rolling and Stirring

做青是形成乌龙茶品质特点的重要工序。将萎调后的叶子置于摇青机中摇动，叶片互相碰撞、摩擦，叶组织被破坏，从而促进酶促氧化作用的进行，这就被称为"摇青"。叶片经过摇动，由软变硬。再静置一段时间，使酶促氧化作用减缓，水分均匀分布，嫩叶恢复弹性，由硬变软。由于多酚类物质从破损细胞中溢出，以及水分的减少，使得叶缘部位氧化反应剧烈，显现出红色物质，形成"红边"，而叶片中央，则由暗绿变为黄绿，构成了乌龙茶特有的"绿叶镶红边"的外形特征，更增加了茶香。

People also name the making green for rolling. It is the most vital process to form the characteristics of Oolong Tea. Put the withering leaves into the stirring machine to stir. It is called "Stirring" Those leaves collide with each other, and leaf tissues break down to promote the oxidation of enzymes. This step is called stirring. The leaves become hard after shaking. Then cool for a while. It can slow down enzymatic oxidation. The distribution of moisture becomes more even. The stretch of the tender leaves is a complete recovery and changes from hard to soft. Due to the polyphenol materials spilling out of the specific damaged cells and the decrease of leaf water exacerbating the oxidation reaction of leaf marginal, the red substance will appear to form the red edge. The middle part of the leaf will change from dark green to yellow-green. It creates the unique physical feature of green leaves with the red side edge of Oolong Tea, and the fragrance is also enhanced.

杀青 Fixation

　　乌龙茶的杀青工艺多采用杀青机进行杀青。杀青时要使茶青能在短时间内达到适宜的温度，以迅速破坏酵素的活性。杀青的时间也要适度掌控，若时间过长，则有可能发酵过度，影响香气的发挥；如时间过短，叶内一些物质转化不能充分进行，会大大影响成品茶的品质。杀青适度的叶子，青涩气消失，香气加浓，水分含量达到揉捻的适度标准。

　　The fixation craft of Oolong Tea is mainly processed with a fixation machine. The optimum temperature of fresh tea leaves should be reached quickly to destruction of enzymes activity active enzymes. The fixation time also needs to be controlled moderately. If the fixation time is very long, it will cause excessive fermentation to influence the fragrance's volatilization. Only a limited leaf conversion is achieved if the fixation time is short. It may significantly affect product quality. Tea leaves are processed under moderate fixation without the green and astringency taste. The aroma will be thicker. The moisture content can reach the appropriate standard for rolling.

　　高温杀青使酶的活性遭到破坏，有效控制了氧化反应的进行，防止红梗红叶的出现，并发散掉低沸点的青涩气，增强了茶的醇香，巩固了茶的品质。

　　High-temperature fixation can destroy the activities of relevant enzymes. It can effectively control the oxidation reaction process to avoid the appearance of red stalks and red leaves. And it can help to spread out the low-boiling green and astringency. It also intensifies the fragrance to consolidate and enhance the quality of the tea.

静置回润
Humidification Step

杀青后的茶叶，在进行下一步的揉捻之前要先用干净的湿布包裹起来，再放入谷斗中，上面覆盖一层湿布，把茶叶略微压实。这一工序对茶叶起到闷热静置的回润作用。

The tea after fixation, wrap up those tea leaves in a clean wet cloth before the subsequent rolling. Then put it into a grain hopper and cover it with a damp cloth layer to slightly compact the tea leaves. This step plays a role in the humidification process of tea.

揉 捻
Rolling

乌龙茶根据揉捻方式的不同，分为散揉和团揉两种。散揉是将杀青后的叶子直接放入揉捻机里压揉；团揉则要先用布把茶青包裹成团，再进行人工或机械的揉捻。揉捻的力度是影响茶叶品质形成的重要因素，力度过大会使叶片易碎，太轻则不利于叶片成形。

The rolling of Oolong Tea can be divided into loose leaves rolling and ball shape rolling by different methods. Loose rolling is the method that put straight into the fixated tea leaves into the rolling machine to press and roll. Ball shape rolling is how you mix fresh tea leaves to form a dough ball and then deal them with artificial rolling or machine rolling. The power of rolling is one of the most critical factors influencing tea quality. The leave is easily broken with too much power. But it isn't easy to shape with too soft power.

干　燥
Drying

乌龙茶的干燥室利用高温来破坏残留的酶的活性，从而彻底抑制发酵反应的进行，充分蒸发水分，可以固定茶的品质。干燥产生的热化反应，能消除茶叶的苦涩味道，发散浓厚的茶香。

The drying room for Oolong Tea is applied at high temperatures to destroy residual enzyme activity. The fermentation has been suppressed altogether. Evaporating the moisture completely can keep the quality of tea. Thermalization processing is produced from drying. It can vanish the bitter and astringency taste to promote a strong tea fragrance.

乌龙茶的干燥方式与绿茶相似，都采用烘笼或机械烘焙。烘笼烘焙在初焙时要经常翻搅，使茶叶干燥均匀，焙至七成干时，要取出摊晾一段时间，使水分重新分布，再进行烘焙。烘焙的时间、温度要视叶子的老嫩程度、含水量、外界湿度等条件灵活掌握。而机械烘焙是在烘干机里进行的，温度、时间都能自动控制，因其方便、快捷、省力，所以是目前茶农最常用的方法。

The drying method of Oolong Tea is similar to Green Tea. The bamboo-frame dryer or machine dryer is applied for them. Bamboo-frame dryer baking is for primary baking. It would be best to stir, dry, and even tea leaves frequently. Take the leaves out to cool for a while when 70% of the tea leaves are dried. It can redistribute the moisture for the next time roasting. The roasting time should be flexible according to the tenderness degree, moisture content, and the leaves external humidity. And the mechanical bakers do it in the dryer. Because baking leaves with a hot air drying machine is the most frequently used method because of its convenience, speed, and labor-saving. The temperature and time can also be controlled automatically.

烹茶四宝
The Four Brewing Treasures Wares

生活在中国闽南、潮汕地区的人们对乌龙茶非常喜爱。品饮乌龙，首重风韵，讲究用小杯慢慢品啜，闻香玩味。冲泡起来也很下功夫，因此称之为饮功夫茶。福建功夫茶历史悠久，自成文化，配有一套精巧玲珑的茶具，美其名曰"烹茶四宝"，指的是潮汕风炉、玉书碨、孟臣罐、若琛瓯。潮汕风炉是一只缩小了的粗陶炭炉，为广东潮汕地区所制，生火专用；玉书碨是一个缩小的瓦陶壶，约能容水20毫升，架在风炉上，烧水专用；孟臣罐是一把比普通茶壶还小的紫砂壶，专作泡功夫茶用；若琛瓯是个只有半个兵乓球大小的白色瓷杯，容水量仅4毫升，通常一套3～5只不等，专供饮功夫茶之用。

Most people, especially those who live in the south Fujian and Chaoshan regions of China, love Oolong Tea. People enjoy the charm. It is also the most important of drinking Oolong Tea. Pat carefully sips the Oolong Tea slowly with a small cup and then smells the fragrance, repeatedly pondering. It would be best if you put much effort into the brewing. So it is called drinking Gong Fu Tea. (Gong Fu here in Chinese means Expend energy). Fujian Gong Fu tea has a long history and independent cultural features. There is also a set of exquisite tea sets with it. It has been gloriously named the four brewing treasures wares. It includes Chaoshan region blast furnace, Yu Shu kettle, Mengchen pot, and Ruochen bowl. The Chaoshan region blast furnace is a smaller rough pottery charcoal furnace in the Chanshan region. It has particular use for lighting up. Yu Shu kettle is a smaller pottery pot with a capacity of twenty milliliters. It is set on the blast furnace for heating water. Mengchen pot is a more miniature ceramic teapot than our usual teapot brewing Gong Fu tea. Ruochen bowl is a white porcelain cup that has only half the size of a Ping-Pong ball. It only has a capacity of 4 milliliters. A set of three to five cups is the most common type. It is unique for the use of brewing Gong Fu tea.

茶具的摆设以孟臣罐为中心，摆放在一个椭圆或圆形的茶盘中，壶、杯、盘可按个人喜好自行搭配，具有独特的艺术价值和艺术美感，缺一不可，往往被看成一套艺术品，为细腻考究的功夫茶艺锦上添花。

You are putting the Mengchen pot in the middle of the tea sets. Putting it into a round shape or elliptical shape tea tray. You can match the teapot, teacup, and tea plate to your favorite taste. It has with unique artistic value and artistic sense of beauty, not a single one can be omitted. It is a set of artwork. It makes exquisite Gong Fu tea with a sense of creative pleasure.

冲泡要领
Brewing Essentials

乌龙茶的冲泡时间由开水温度、茶叶老嫩和用茶量多少这三个因素决定。一般情况下，冲入开水2~3分钟后即可饮用。但是，有下面两种情况要做特殊处理：一是如果水温较高，茶叶较嫩或用茶量较多，冲第一道可随即倒出茶汤，第二道冲泡后半分钟即可倾倒出来，以后每道可稍微延长数十秒时间；二是如果水温不高、茶叶较粗老或用茶量较少，冲泡时间可稍加延长，但是不能浸泡过久，要不然汤色变暗，香气散失，有闷味，而且部分有效成分被破坏，无用成分被浸出，会增加苦涩味或其他不良气味，茶汤品味降低，若是泡的时间太短，茶叶香味则出不来。乌龙茶较耐泡，一般可泡饮5~6次，上等乌龙茶更是号称"七泡有余香"。

The brewing time of Oolong Tea is determined by three factors that the temperature of the water, the tender degree of tea leaves, and the tea amount. They are generally brewed into boiling water for two to three minutes to drink. But these below two situations need to be dealt with specially. One is brewing more tender tea leaves or more tea amount with higher temperature water. Pouring the first infusion at once. Pouring the second infusion after brewing for half a minute. Slightly prolong the time of seconds for each infusion after the second infusion. The second situation is brewing with low-temperature water, tea leaves are course, and the tea amount is less. The brewing time can be extended slightly. But it cannot be soaked for a long time. Or the tea soup will be darker, and the fragrance will disappear. It will also with stewed taste. Some active components will be damaged. The useless part will be soaked to enhance the bitter taste or other bad smells. The grade and taste of tea will be lower. Tea fragrance cannot soak out with the short brewing time. We can brew the Oolong Tea many times. The first-class Oolong Tea, also called it, still has fragrance after seven times brewing.

红茶
Black Tea

红茶具有红叶、红汤的外观特征，色泽明亮鲜艳，味道香甜甘醇。红茶中含有丰富的蛋白质，保健作用极好，其性甘温，可养人体阳气，生热暖腹，温胃祛寒，消食开胃，增强人体的抗寒能力。适宜脾胃虚弱者、体质偏寒者饮用。

虽然红茶中所含的酚类成分与绿茶相比有较大的区别，但红茶同样具有抗氧化、降低血脂、抑制动脉硬化、杀菌消炎、增强毛细血管功能等功效。

With the appearance characteristics of red leaf and red tea soup of Black Tea, the tea soup is brilliant with a sweet and fresh taste. Black Tea is very rich in protein. Therefore, it is suitable for health care. The property is warm in Chinese medicine to refresh Yang energies. It generates heat to warm the abdomen and stomach to dispel the cold. It can promote good digestion and improve cold resistance. It is best for those with spleen and stomach weakness and kidney-Yang deficiency. There is a noticeable difference in the phenolic compounds between Green Tea and Black Tea. But Black Tea also has antioxidative, decreasing the blood-lipid and inhibiting atherosclerosis: antiseptic, antiphlogistic effect and enhanced reabsorption of the capillary vessel.

红茶的品质
Quality of Black Tea

小种红茶
Xiaozhong Black Tea（Souchong Black Tea）

　　小种红茶为中国福建省特产，有正山小种和外山小种之分。正山小种产于风光秀美的福建武夷山地区。"正山"乃是真正的"高山地区所产"之意。凡是武夷山中所产的茶，均被称作正山。武夷山附加地区所产的红茶均为仿照正山品质的小种红茶，质地较为逊色，统称外山小种。

　　Xiaozhong Black Tea（Souchong Black Tea） is the specialty of Fujian Province. It can be divided into the Lapsang Xiaozhong Tea and other areas of Xiaozhong. The meaning of Lapsang is that the masters make tea in mountainous regions. All the tea grown in the Wuyi Mountains is called Lapsang. Xiaozhong tea from other areas is Black Tea of inferior quality. Therefore, they are counterfeited o be Lapsang Xiaozhong Tea.

正山小种条索饱满，色泽乌润，泡水后汤色鲜艳绚丽，香气绵长，滋味醇厚，具有天然的桂圆味及特有的松烟香。由于其优越的自然生长环境，还具有独特的保健功效，长期饮用可保健养身。正山小种迄今已有四百余年的历史，是世界上最早出现的红茶，早在17世纪初就远销欧洲，并大受欢迎，成为欧洲人心目中中国茶的象征。

The appearance of Lapsang Xiaozhong Tea is black, blooming in full shape. The tea soup is red brilliant with a lingering aroma and has a mellow and thick taste. It also has natural cinnamon and a smoky pine flavor. The finished tea is given some extra drying over a smoking pine fire, imparting a sweet, clean smoky flavor to the tea. Because of its excellent growing environment, it also has unique health benefits. Long-term drinking is good for health care. Lapsang Xiaozhong Tea has a long history of close to 400 years. It was the first kind of Black Tea in the world. In the early 17th century, it was exported to Europe and earned much favor from consumers. It has been the symbol of Chinese tea.

茶类篇 The Category of Tea

图片由茶人梁骏德提供
Photos by courtesy of Liang Junde

功夫红茶
Gongfu Black Tea

功夫红茶又名条红茶，经过萎凋、揉捻、发酵和干燥的流程制成，是中国特产的红茶品种。因其工艺高超，制作精细，品饮讲究而得名。根据茶树品种又分为大叶功夫茶和小叶功夫茶。大叶功夫茶是以乔木或半乔木茶树鲜叶制成；小叶功夫茶是以灌木型小叶种茶树鲜叶为原料制成。

Gongfu Black Tea is a kind of rolled strips and curls of Black Tea, processed by withering, rolling, fermentation, and drying procedures. It is a unique variety of Chinese Black Tea and is famous for its superb crafts, delicate production and delicate drinking methods. It can be divided into big-leaf Gongfu tea and small-leaf Gongfu tea. The big-leaf Gongfu tea is made of fresh leaves from deep-rooted tea trees. The small leaf Gongfu tea is made of fresh leaves from the shrub tea trees.

功夫红茶条索挺秀，紧细圆直，香气鲜浓纯正，滋味醇和隽永，汤色红明，叶底红亮。中国功夫红茶品类多，产地广，有12个省先后生产出功夫红茶，按产地不同，品质各具特色。

The appearance of Gongfu Black Tea is tender and straight. It is also tight, slender, and round. The aroma is fresh and pure, with a mellow and meaningful taste. The color is clear red. The brewed leave is bright red.

There are lots of categories of Gongfu Black Tea in China. It is more widely. Twelve provinces successively produce the Gongfu Black Tea. Each one has its characteristics according to the different producing areas.

茶类篇

The Category of Tea

红碎茶
Broken Black Tea

红碎茶有百余年的产制历史，是国际市场上销售量最大的茶类。它是在功夫红茶加工技术的基础上，以揉切代替揉捻，或揉捻后再揉切制成。揉切的目的是充分破坏叶组织，使干茶中的内含成分更易冲泡出，形成红碎茶汤色红艳明亮，滋味浓、强、鲜的品质风格。根据其总的品质特征，红碎茶可分为叶茶、碎茶、片茶、末茶四个细类。中国云南、两广和海南地区是红碎茶的集中生产地。国外红碎茶的生产主要集中在印度、斯里兰卡和肯尼亚，其产量的总和占世界红碎茶总产量的80%以上，且质优价高。

Broken Black Tea has over one hundred years of producing history. It is the biggest-selling consumer variety in the international market. The twisting and cutting replaced the step of wisting or twisting after twisting and then cutting based on processing technology on Gongfu Black Tea. The purpose of twisting and cutting is to destroy the leaf tissue fully, so that, the dry tea's conterits are easier to be brewed out. Form the red brilliant and clear tea liquor color with the characteristics of a strong, mellow, and fresh taste. According to its overall quality characteristics, Broken Black Tea is divided into four small cate gories, leaf tea, broken tea, piece tea, and powdered tea. The Yunnan areas, Guangdong, Guangxi provinces and Hainan regions are broken Black Tea concentrated production centers. Overseas Broken Black Tea is mainly produced in India, Sri Lanka, and

Kenya. The total output amounted to more than 80% of the worldwide production. It is a high quality high price premium Black Tea.

名茶种类
The Variety of Chinese Famous Black Tea

祁门功夫红茶是中国传统功夫红茶的珍品,有百余年的生产历史。主产于安徽省祁门县,简称"祁红"。祁红功夫茶条索紧秀细长,色泽乌黑泛灰光,俗称"宝光",内质香气浓郁高长,清雅隽丽,似蜜糖香味,极品茶更是蕴含兰花香气,清鲜持久,号称"祁门香"。清饮最能体味祁红的隽永香气,即使添加鲜奶亦不失其醇香。春天饮红茶以它最为适宜,作为下午茶、睡前茶也很适合。

Keemun Black Tea is a treasure of Chinese traditional Gongfu Black Tea with hundreds of years of history. It is mainly produced in Keemun County in Anhui Province. So it is known simply as Qi Black Tea. Tea leaf is slight and slender. The color is dark grey. The tea is elegant, with a strong flavor and honey-sugar aroma. The superior Keemun Black Tea also has a fresh and lasting orchid fragrance. Hence, it is known as the Keemun flavor. You can fully appreciate the long-lasting fragrance in light drinking. However, it will not lose its mellow and sweet taste, even after adding fresh milk. Spring is the best season for drinking this kind of tea. It is also suitable as an afternoon tea and before-bedtime tea.

品级的划分
Grade Division

红茶的品级根据品种、采摘部位、产区、海拔高度及季节等而有所不同，很难只凭其中某一项标准来界定品级。世界上红茶的品种很多，产地也很广泛。其中最负盛名的四大名品红茶有：祁门红茶、阿萨姆红茶、大吉岭红茶和锡兰高地红茶。

The different varieties decide the grade of Black Tea, other parts of picking, producing area, altitude, and seasons. Therefore, it is hard to grade it by one of the standards. There are varieties of Black Tea globally, and the origins of Black Tea are quite extensive. There are four famous varieties: Keemun Black Tea, Assam Black Tea, Darjeeling Black Tea, and Ceylon Black Tea.

红茶的制作
Black Tea Processing Techniques

发酵茶
Fermented Tea

红茶属于全发酵茶,因此发酵也是红茶制作中最重要的工序,也是与制作其他茶叶最显著的区别。

The Black Tea is fully fermented. Fermentation is the most critical processing step in making Black Tea. It is also the most apparent difference between other varieties of tea.

中国的红茶种类主要有功夫茶、红碎茶和小种红茶三种,其主要制作工序都有萎凋、揉捻、发酵、干燥四个步骤,但各道工序需要的条件和程度又略有不同。下面以功夫红茶为例,对红茶制作的主要步骤作逐一介绍。

There are three main types of Chinese Black Tea: Gongfu Black

Tea, broken Black Tea, and Souchong Black Tea. The central processing procedures are the same. There are withering, rolling, fermenting, and drying. But the conditions and degree are slightly different, which are required by each step. Now we take the Gongfu Black Tea to introduce the main phases of Black Tea processing.

日光萎凋
Sunlight Withering

萎凋是红茶加工的第一道工序。红茶萎凋有三种方法：日光萎凋、室内自然萎凋和萎凋槽萎凋。日光萎凋这种方法受天气制约很大，阳光强烈的午后和阴雨的天气都不适宜。通常只在春茶季节，气候比较温和时被采用，这个时节萎凋程度容易控制，萎凋时间大约需要一小时。

Withering is the first step in Black Tea processing. There are three ways of Black Tea withering: sunlight withering, indoor natural withering, and trough withering. The natural conditions and weather constrain the sunlight withering among these. For example, the afternoon in the intense sun and the rainy weather may not be appropriate. Usually, this method is used in the warm season, such as in spring tea. The degree of withering can be easily controlled at this time of year. The withering time is about one hour.

揉 捻
Rolling

揉捻是红茶加工的第二道工序。揉捻使叶细胞遭到破坏，叶卷成条，叶汁溢出并凝于叶表，增加了茶叶的浓香，为发酵创造条件。揉捻需要的空气相对湿度为85℃~95℃，室内温度保持在20℃~24℃的条件下进行，需要避免日光直射。在夏秋季节，低湿高温的环境下，也可通过安装喷雾、洒水、搭荫棚等来降低温度、提高湿度。揉捻时间和萎凋叶的投入量根据茶树品种、揉捻机型号而定。

Rolling is the second step for Black Tea manufacture, which damaging the leaf cells. Tea leaves are rolled up into strips. The leaf juice might spill over to freeze on the leaf surface. The aroma can be incredibly stronger and more fragrant. It creates conditions for fermenting. The air relative humidity for rolling is about 85℃ ~95℃ . The inside temperature should be remained between 20 and 24 degrees Celsius and be kept out of the sun. Installing sprays, sprinklers, and shade shed can help lower the temperature and increase the humidity under the condition of low humidity and high–temperature surroundings during the summer and autumn. The rolling time and tea amount for withering are according to tea tree varieties and rolling machine types.

发 酵
Fermenting

发酵是红茶加工中最关键的工序。它使氧化酶的活性增加，与多酚类物质发生氧化聚合，叶子变为红色。发酵室要求相对湿度达 95% 以上，温度一般在 22℃~25℃。发酵时将揉捻叶平铺在特定的发酵盘中，嫩叶稍薄，老叶略厚；春茶需薄铺，夏秋茶略厚铺。

Fermenting is the most critical stage in the Black Tea production process. It can increase the activity of the oxidizing enzymes. As a result, the leaves turn red because of oxidation polymerization. The indoor relative humidity should be above 95%, and the temperature should be controlled between 22℃ and 25℃. Those rolling leaves are laid on the fermentation plate for fermentation. The tender leaves are slightly thin in Spring, and older leaves are somewhat thick in Summer and Autumn.

干 燥
Drying

干燥是红茶制作的最后一道工序。它是通过高温来达到钝化酶的活性，使发酵停止，同时蒸发水分，固定茶型，防止霉变。红茶一般要经毛火和足火两次干燥。毛火干燥时，需高温烘焙，薄薄摊铺；然后再足火干燥，此时温度应稍低，摊铺微厚，时间较毛火略长，直至含水量小于6%。毛火干燥适度的叶子，用手触摸会有柔软、刺手、有弹性的感觉。足火后干燥程序基本完成，茶叶若用力手捻则成粉末状，茶色更重，茶香更浓。

Drying is the last step in the Black Tea production process. The high temperature can inactivate the enzymes. The fermentation will stop. At the same time, it can evaporate moisture to firm its shape. It also can be prevented from going moldy. Usually, the Black Tea needs to be dried twice for the first and final drying. Baking tea at high temperature, then tedding thinly during the first drying. Then process the tea with final drying. The temperature should be slightly lower for this process. Tedding tea leaves should be slightly thick for a longer time than the first drying. The moisture content should be lower than 6%. You can feel soft, spiny, and elastic when touching the tea leaves if the first drying is applicable. The drying procedure is completed after the final drying. The tea soup color will be deeper, and the tea aroma will be more assertive with the powder tea.

红碎茶的揉切
Rolling and Cutting of Broken Black Tea

红碎茶与功夫红茶在制法上最大的区别就在于揉切。嫩度较差的叶子，一般先用揉切机进行揉捻后再揉切。对于较嫩的鲜叶可将萎凋后的叶子直接放入揉切机里进行揉切，红碎茶的外形条件也因此而形成。

The most significant difference between broken Black Tea processing and Gongfu Black Tea processing is the step of rolling and cutting. We usually roll and cut tea leaves with Rolling and Cutting machine. Then directly put the tea leaves into the machine after withering. The appearance of Black Tea has already been created with these steps.

揉切过程中，叶子受到多种力的作用，温度迅速升高，为避免叶温过高而引起过度的发酵，通常要缩短揉切时间，但为了保证碎茶的效果，则要增加揉切的次数。由于叶片被切碎，使得叶细胞遭到严重破坏，叶汁外溢，叶内所含物质与空气充分接触，氧化作用加剧，由此便形成了红碎茶香气馥郁，口感浓醇的特点。

The tea leaves are under various forces in rolling and cutting. The temperature rises rapidly. The time of rolling and cutting should be shortened to avoid too high temperature would cause excessive fermentation. But to ensure the excellent quality of broken Black Tea and frequency of rolling, cutting must be stepped up. Due to tea leaves being cut and leaf cells being seriously destroyed to spill juice overleaf, leaf substances could make the most contact with the air to get the best oxidation and fermentation. And these steps makes the fragrance aroma with the rich and mellow taste of Broken Black Tea.

小种红茶的干燥
Drying of Xiaozhong Black Tea

小种红茶不同于功夫红茶的制作工艺之处，在于萎凋和干燥过程中加入了松烟烘焙。其主要方法是利用松柴燃烧产生热量来蒸发多余水分，同时茶叶吸收掉大量的松烟，促进芳香物质的散发，形成小种红茶所具有的烟熏香味，以及口感醇正浓厚的品质特点。

There are some differences in the processing technique between Xiaozhong Black Tea and Gongfu Black Tea. There is one more processing step. They are burning turpentine soot to bake in the processes of withering and drying. The primary method is to use the heat from pine wood burning to evaporate excessive moisture. Tea leaves absorb a lot of pine smoke to give off aromatic substances. So the Xiaozhong Black Tea will also have a smoke smell with a mellow, regular, and heavy taste.

红茶的冲泡
The Brewing of Black Tea

适宜的茶具
Appropriate Tea Sets

红茶高雅的芬芳及香醇的味道，必须要以合适的茶具搭配，才能烘托出它独特的风味。品饮红茶最合适的茶具是白色瓷杯或瓷壶，尤以骨瓷最佳。质地莹白、隐隐透光的骨瓷杯盛入色彩红艳瑰丽的红茶茶汤，在升腾的雾霭中感受扑鼻而来的香气。闲暇时捧着一杯红茶，悠然度过一个轻松的午后，保温性能最佳的骨瓷杯能保证你品到的每一口茶都温暖且甘甜。

The elegant fragrance and mellow taste of Black Tea should be brewed with the appropriate tea sets to express its unique flavor. The best tea set for tasting Black Tea is a white porcelain cup or porcelain pot, and especially the bone china cup is the best. The pale white color with a low light transmission teacup can set off the brilliant red tea liquid color. The nose is bursting with the fragrance of Black Tea in wave rising aroma of the cup. We can spend a pleasantly lazy afternoon with a cup of Black Tea. The bone china cup, which has the best heat preservation performance, can ensure you thoroughly enjoy each mouthful's warm and sweet flavor.

一般来说，功夫红茶、小种红茶、袋泡红茶、速溶红茶等大多采用杯饮法，即置茶于白瓷杯中，用沸水冲泡后品饮。红碎茶和片末红茶则多采用壶饮法，即把茶叶放入壶中，冲泡后使茶渣和茶汤分离，从壶中慢慢倒出茶汤，分置于各小茶杯中，便于饮用。茶叶残渣仍留在壶内，或再次冲泡，或弃去重泡，处理起来都很方便。

Generally speaking, Gongfu Black Tea, Souchong Black Tea, bagged Black Tea, and instant Black Tea is mainly used cup-drinking method. It is how to put tea leaves into a bone china cup, then brew it with boiling water for drinking. The tea dregs and tea liquid will be separated after brewing. Pour the tea soup slowly and divide it into several teacups for drinking. Then tea dregs are still in the pot. You can choose to brew it one more time or throw it. It is very convenient.

清饮法
Light Drinking (Pure Drinking)

清饮是指将茶叶放入茶壶中,加沸水冲泡,然后注入茶杯中细品慢饮,不在茶汤中加入任何调味品,体味的完全是红茶固有的芬芳。苏东坡曾有诗比喻——"从来佳茗似佳人"。清饮的红茶,正如一位天生丽质的美人,不需要人工的雕饰,也能散发出自然的韵味。

Light Drinking is how we put the tea leaves into a teapot and then brew them with boiled water. Taste and enjoy the tea soup with a cup of tea. Tea can display the natural fragrance of Black Tea without any additional flavors. In his poem, the famous poet of the Song Dynasty, Su Shi, imagined the beautiful woman as an excellent tea. Black Tea with light Drinking is just regarded as a natural beauty. They give off a natural aroma without artificial carved.

清饮时,一杯好茶在手,慢慢啜饮,默默赏味,最能使人进入一种忘我的精神境界,欢愉、轻快、激动、舒畅之情油然而生。中国人多喜欢清饮,特别是名特优茶,一定要清饮才能领略其独特风味,享受到饮茶乐趣。

Enjoying a cup of good tea with a light drinking method can help experience a state of no thought. The feeling of joy, lightness, excitement, and comfort could arise spontaneously. Chinese people prefer light drinking, especially for best-grade tea. You can enjoy the unique flavor and the pleasure of light drinking.

调饮法
Tea Drinking with Flavoring

既然佳茗堪比佳人，那自然是浓妆淡抹总相宜。除去中国人传统的清饮法外，受西方人影响，现在美味丰富的调饮法也同样流行。调饮法，是在泡好的茶汤中加入奶、糖、柠檬汁、蜂蜜、咖啡、香槟酒等，以佐汤味。所加调料的种类和数量，根据个人爱好，任意选择调配，风味各异。也有的在茶汤中同时加入糖和柠檬汁或蜂蜜和酒同饮，或将其置于冰箱中制成不同滋味的清凉饮料等，都别具风味。

Since excellent tea is often compared to natural beauty, tea can be drunk lightly or in multiple ways. Unlike traditional Chinese light drinking, delicious and colorful flavoring drinking is also prevalent. Tea Drinking with Flavoring adds milk or sugar, lemon juice, honey, coffee, champagne, etc. A lot of it has to do with personal preference or taste. Someone also drinks by adding sugar with lemon and honey to wine. Or it can be kept cool in refrigerators to make different tastes of cooling drinking. It must be a unique experience.

调饮法在现代广为流行,尤其受到年轻人的喜爱。调饮法用的红茶多数是用红碎茶制的袋泡茶,茶汁浸出速度快,浓度大,也易去茶渣。

Flavoring drinking tea is widely popular in modern life, especially oved by young people. The Black Tea used for flavoring drinking is mostly tea bags made from Broken Black Tea. The leaching rate is high, and the tea is thicker. It is also easier to remove tea residues.

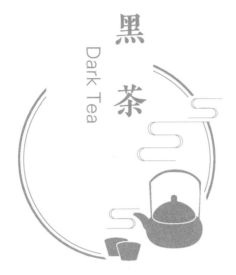

黑茶
Dark Tea

茶类篇 The Category of Tea

黑茶流行于云南、四川、广西等地，同时也深受藏族、蒙古族和维吾尔族同胞们的喜爱，几乎已经成为他们日常生活中的必需品。黑茶呈黑色，汤色近似深红，叶底匀展乌亮。

Dark Tea is popular in Yunnan, Sichuan, Guangxi and other places. And it is trendy among the Tibetan, Mongolian and Uighur compatriots. It has almost become a necessity in their daily life. The Dark Teacake is black. The tea soup is deep red with a glossy black and open brewed leaf.

对于喝惯了清淡茶叶的人，初尝味道偏苦，口感浓醇的黑茶或许觉得难以下咽。但只要坚持长时间饮用，很多人都会爱上它独特的"滑、醇、柔、稠"的滋味。

It will be a big challenge for those who are used to drinking light tea. There is a feeling of a bitter taste for the first time. The heavy and mellow Dark Tea is hard to swallow. But many people will love its unique flavor of smooth, mellow, soft and thick if they drink it for a long time.

　　黑茶在发酵的过程中产生了一种普诺尔成分,可以有效防止脂肪堆积,抑制腹部脂肪增加。所以近年来黑茶在社会上被推崇甚广。

　　Produces one Pu Nuoer component in the fermentative process, preventing the role's fat stack inhibit the increase of abdominal fat. So Dark Tea has been rridely distributed worldwide.

黑茶的品质
Quality of Dark Tea

湖南黑茶
Hunan Province Dark Tea

湖南黑茶原产于湖南安化，现在已扩大到周边的益阳、汉寿等地区。黑茶鲜叶采摘以新梢青梗为主要原料，但不采一芽一二叶。黑茶可分为四个级别：一级以一芽三四叶为主，条索紧卷、圆直，叶质较嫩，色泽黑润；二级以一芽四五叶为主，条索尚紧，色泽黑褐尚润；三级以一芽五六叶为主，条索欠紧，呈泥鳅条，色泽纯净呈竹叶青，带紫油色或柳青色；四级以对夹驻梢为主，叶张宽大粗老，条索松扁皱折，色黄褐。

Hunan Province Dark Tea is native to Anhua County, Hunan Province. It has now expanded to the Yiyang, Hanshouand other surrounding areas. The primary material for producing this kind of Dark Tea is the green stalk with newborn shoots. Do not pick up one bud with one or two tea leaves. It can be divided into four primary levels. One bud with three or four leaves is mainly for the first-grade Hunan Province Dark Tea material. The second-grade material is one bud with four or five leaves, and the dry tea is approached tight with the color of black auburn and bloom. One bud with five or six leaves is in the three grade. The dry tea is less tight to form the loach shape. The color is clear, purple oil and willow green. The fourth grade is mainly with leaves of the top growing shoot. The leaf is open, broad, coarse, and aged. It is loose, flat, wrinkled, and yellowish auburn.

湖南黑茶的制造工艺包括杀青、初揉、渥堆、复揉、干燥五道工序，经蒸压装篓后称为"天尖"，蒸压成砖形的是黑砖、花砖或茯砖等。高档茶较细嫩，低档茶较粗老。茶汤滋味浓醇，无粗涩味，具有松烟香。

图片由茶人肖益平提供
Photos by courtesy of Xiao Yiping

The manufacturing process of Hunan Province Dark Tea includes five steps: fixation, primary rolling, piling, re-rolling, and drying. First, the tea of Tianjian Grade is autoclaved to put into a bamboo basket. Autoclaved to form the brick shape are black brick tea, decorative pattern brick tea, and Fu brick Dark Tea. The higher-grade tea is much tenderer, and the low-grade tea is coarse. The tea soup is mellow and strong without astringency. It has a unique pine-smoky flavor.

四川边茶
Sichuan Province Border-Sale Tea

四川边茶生产历史悠久，分"南路边茶"和"西路边茶"两类。清乾隆年间，规定雅安、天全、荣经等地所产的边茶专销康藏，属南路边茶；灌县、崇庆、大邑等地所产边茶专销川西北松潘、理县等地，称西路边茶。

With a long history, the Sichuan Province Border-Sale Tea industry produces. It is divided into South Border-Sale Tea and West Border-Sale tea. The government set the regulation that the tea made in Yaan, Tianquan, and Rongjing County could only be sold to Tibetan areas during the Qianlong era of the Qing Dynasty. It was called South Border-Sale Tea. The West Border-Sale tea grew in Guan County, Chongqing, and Dayi and is sold to Songpan, Lixian, and other Northwest China places.

南路边茶的原料是采摘当季或当年成熟新梢枝叶，杀青之后经过多次"渥堆"晒干而成。成品茶品质优良，经熬耐泡，是压制"康砖"和"金尖"的原料，最适合以清茶、奶茶、酥油茶等方式饮用，深受藏族人民的喜爱。

The mature tea leaves with a newborn shoot of that season were the law material tea-producing South Border-Sale Tea. It was processed several times, pilling and drying after fixation. The quality of finished tea is good for several times, brewing and soaking. It is the material for compressing the Kang Zhuan Brick Tea and Jin Jian Tea. Suitable drinking methods are pure drinking, adding milk, and buttered tea. The people of Tibet deeply love it.

将当年或1~2年生茶树枝叶采割杀青后直接晒干即成西路边茶。西路边茶的鲜叶原料比南路边茶更粗更老。西路边茶色泽枯黄，是压制方包茶的原料。制造茯砖茶的原料茶含梗量约为20%，而制造方包茶的原料茶则更粗老。

The West Border-Sale tea is processed with newborn tea leaves or 1-2 years of tea leaves. The leaves are dried after fixation. Fresh tea is coarser than South Border-Sale tea. The West Border-Sale tea is dry yellow. It is the material for compressing Fangbao Tea. The raw tea used to make Fu Bricks contains about 20% of tea stalks. The tea material for compressing the Fangbao tea is the coarsest.

滇桂黑茶
Yunnan and Guangxi Provinces Dark Tea

滇桂黑茶顾名思义，是生长在云南和广西的黑茶的统称，属特种黑茶。其品质独特，香味以陈为贵，在我国港澳地区和东南亚及日本等地有广泛的市场。

As the name suggests, Yunnan and Guangxi Provinces Dark Tea is a particular type of tea that grows in those areas. The quality is unique. It is precious for its stale flavor. It has a broad market in Hong Kong, Australia, Southeast Asia, and Japan.

云南黑茶是用滇晒青毛茶经潮水渥堆发酵后干燥制成。这种茶条索肥壮，汤色明亮，香味醇浓，带有特殊的陈香，可直接饮用。以这种茶为原料，可蒸压成不同形状的紧压茶——饼茶、紧茶、圆茶等。

The fresh primary tea is processed with pile-fermentation and drying for the Yunnan Province's Dark Tea. The tea leaf is bold with bright and mellow tea soup. It also has a unique aged scent. It can be drunk directly. Use this primary tea as the material. It can be autoclaved to form different compressed tea shapes such as tea cake, tight tea, and round tea.

广西黑茶最著名的是六堡茶，已有 200 多年的生产历史。六堡茶制作工艺流程是杀青、揉捻、渥堆、复揉、干燥，制成毛茶后再加工时仍需潮水渥堆，蒸压装篓，堆放陈化，最后使六堡茶的汤味形成红、浓、醇、陈的特点。

The most famous type of Guangxi Province Dark Tea is Liu Bao Tea. which has more than two hundred producing histories. The production process includes Fixation, Rolling. Piling, Re-rolling, Drying. It still needs to be processed with wet pile-fermentation and then autoclaved to put into a bamboo basket and pile up for aging when reprocessed. Finally, it can form red, heavy, mellow, aged tea soup of Liu Bao Tea.

紧压茶
Compressed Tea

将黑毛茶、老青茶及其他适制毛茶经过高温、高湿与压力蒸压的方式加工成饼形、砖形、团形等状态的茶叶，称之"紧压茶"，主要销往边疆少数民族地区。紧压茶根据堆积、做色方式的不同，分为"湿坯堆积做色""干坯堆积做色""成茶堆积做色"等种类。多数品种配用的原料比较粗老，风味独特，具有减肥、美容等效果。

As usual, primary dark tea, aged Green Tea making pressed tea and other kinds of primary tea suitable for manufacturing must be processed at high temperatures, high humidity, and high pressure. It can be processed with autoclaves to form pie, brick, mass, and other shapes. Hence, it is called compressed tea. They are mainly sold to the frontier ethnic regions of the country. Based on The different piling-up and color-making methods, compressed tea can be divided into water body piling up to make the color, dry body piling up, finished tea piling up, and other varieties. Most materials are coarse with a unique taste, which has the effect of losing weight, beauty, and so on.

中国紧压茶产区比较集中，主要有湖南、湖北、四川、云南、贵州等省。目前，中国生产的紧压茶大多数为砖茶。由于砖茶与散茶相比更为紧实，所以用开水冲泡难以浸出茶汁，饮用时必须先将砖茶捣碎，在铁锅或铝壶内煎煮才可以饮用。

The production areas are mainly distributed in Hunan, Hubei, Sichuan, Yunnan, Guizhou, and other provinces. So far, most of the compressed tea produced in China is brick tea. Brick tea is different from bulk tea. It is incredibly tight and heavy. So it is hard to soak out tea with boiling water. Instead, you need to mash the brick tea into a piece to cook in an iron or aluminum pot for drinking.

名茶种类
The Variety of Chinese Famous Dark Tea

普洱茶是以云南省的云南大叶种晒青毛茶为原料，经过后发酵加工成的散茶和紧压茶，是历史悠久的云南特有的地方名茶。普洱外形色泽褐红，内质汤色红浓明亮，香气独特陈香，滋味醇厚回甘，叶底褐红。新普洱茶味道浓烈，刺激性强，而老的普洱茶由于陈放较久，能持续进行自然发酵过程，茶性变得温和无刺激，存放的时间越久，氧化程度越高，茶汤滋味越醇厚，能促进血液的新陈代谢，不刺激肠胃，还能养生、助气、补气，甚至还有降血脂、瘦身、抗癌等功效。

The material for Pu Er Tea is the big leaf primary tea with sunlight withering in Yunnan Province. Bulk Tea and Compressed Tea with post-fermented is a unique and famous tea in Yunnan Province with a long history. Pu Er Tea is brown-red with bright red tea soup and a unique aged scent. Its taste is mellow and sweet after the taste. The brewed tea is brown-red. The new tea's taste is much stronger with sharpness. Due to being stored up for a long time, the natural fermentation is still ongoing for the aged Pu Er tea. The character is mild, with no irritation. With longer storage time, the degree of oxidation is higher for Dark Tea. The mellow and thick tea soup can promote the metabolism of blood. There is no stomach irritation, but it can also keep healthy and Tonifying Qi in Chinese medicine. It even lowers blood fat, reduces body weight, and reduces anticancer.

因加工方法有所不同，普洱茶在严格意义上来说并不属于黑茶，但通常被人们归入黑茶种类。作为一种健康饮品，近年来普洱茶开始在全国广泛流行，形成一股普洱热潮，成为黑茶类中的典型代表。由于其在各类茶叶中独特的越陈越香，越陈功能越显著等特点，使得普洱茶升华为茶叶中具收藏鉴赏价值的"古董"，如储存保管得当，可储存多年仍能保持其原有风味。其市场价值也随年份一路飙升，蔚为可观。

Due to different processing methods, Pu Er Tea is not strictly dark tea. But it is usually included in Dark Tea. As a health drink, Pu Er Tea has been widely famous throughout the country recently and has become enthusiastic. Pu Er Tea has become the most famous and typical representative of Dark Tea. Due to its unique characteristics of getting better with age, the Pu Er Tea is an antique with appreciation value among teas. It can keep its original flavor when stored and used correctly. Therefore, its market value has been on the up for years.

黑茶的制作
Dark Tea Processing Techniques

杀青
Fixation

由于黑茶的鲜叶粗老，含水量低，杀青前要先对鲜叶进行洒水处理，利用水分受热形成蒸汽来提高叶表温度，从而达到杀匀杀透的目的。

Due to the fresh leaves being coarse and having low moisture content, we need to water the fresh tea leaves before fixation. The steam is formed by heating the moisture to increase the leaf surface temperature, which can help to achieve the goals of even and enough fixation.

黑茶杀青分手工杀青和机械杀青两种。手工杀青采取高温快炒的方式，通常选用大口径铁锅，呈30°角倾斜装置在灶台上，每次

168

投放 4~5 千克鲜叶，双手快速翻炒至烫手，再换用三叉状的炒茶叉斗炒，这就通常所说的"亮叉"。待出现大量水蒸气后，双手执叉，转滚闷炒，俗称"握叉"。机械杀青与绿茶大致相同，区别在于当锅温达到要求时，先进行闷炒，再透炒，如此交替进行，至杀青适度方可。黑茶杀青使叶子变为暗绿色，青气消失，叶梗叶片也随之变得柔软。

The fixation methods for Dark Tea can be divided into two ways manual fixation and machine fixation. Usually, use the cauldron to stir fry the fresh tea leaves at high temperatures. Put the cauldron on the kitchen table at 30 angles. Put four to five grams of fresh tea leaves every time. Stir rapidly with your hands until you feel the leaves turn soft and hot. Then stir fry tea leaves with a trident fork, known as the bright fork. Take the fork with the hands to stir fry nightly tea leaves after a large amount of water vapor appears. It is commonly known as Hold the Folk. The machine fixation for Dark Tea is about the same as Green Tea. The difference is that the Dark Tea leaves need to be stirred enough to reach moderate fixation when the temperature meets the process requirement. Tea leaves of Dark Tea will turn to be dark green with fixation. The green odor of tea will be gone, and the tea stalk and leaves will be soft and tender.

初 揉
Primary Rolling

杀青叶出锅后，为避免水溶性物质随水分蒸发和热量散失而凝固，叶片变硬，不利于外形塑造及叶细胞的破坏，应立即趁热放入揉捻机里揉捻。要遵循"轻压、慢揉、短时"的原则。每分钟40转为宜，叶温保持在50℃~60℃左右。揉捻适度的嫩叶卷曲成条，老叶出现褶皱，叶汁附于表面，散发出淡淡的茶香。

Avoiding water-soluble substances will solidify with the evaporation of water and heat loss. Tea leaves will turn harder and bad for the shape of leaf cells' appearance and destruction. Tea leaves need to be rolled into a rolling machine while warm. The step must follow light pressure, slow rolls, and short time. The suitable speed of the device is 40 rotations per minute. The temperature of leaves should be controlled at around 50 ℃ ~60 ℃. The tender leaves will be rolled into strips with moderate rolling, and aged leaves will appear in folds. Tea juice will be attached to the surface and secret the unique flavor of the tea.

渥 堆
Piling

渥堆是形成黑茶独特品质的关键工艺。渥堆要在洁净、无阳光直射的环境下进行，室温一般在25℃以上，相对湿度控制在85%左右。将揉捻后的叶子堆积起来（通常一二级的叶子须解块，三四级的叶子无须解块），覆盖上湿布，以达到保湿保温的目的。中间要适时翻动一次。当茶坯表面出现热气凝结的水珠，发出浓烈的酒糟气味时，青气消失，叶色由暗绿变为黄褐色，茶团黏性减少，容易打散，则渥堆方为适度。渥堆过程中，茶叶内含物发生了一系列的化学反应，使黑茶的口感醇而不涩。

The piling is the critical process step to form the unique quality of Dark Tea. Piling should be processed in a clean environment without direct sunlight. The indoor temperature should be above 25 ℃ with a relative humidity of 85%. Pile the tea leaves after rolling and cover a damp cloth to preserve heat and moisture. It needs to be stirred at the right time during this process. When the water drops appear on the surface of the tea dhool and create a strong slots smell, the green odor will lose. Dark Green Tea leaves will also change to yellowish-brown. This is because the stickiness of solid tea is less and easier to break up. Then the appropriate point of piling is to achieve. An array of chemical reactions are complicated between tea inclusions to make the taste of Dark Tea mellow and without astringent.

干　燥
Drying

黑茶干燥一般采用烘焙法，是在"七星灶"上用旺火烘焙的，达到适宜温度时摊铺第一层茶坯，烘至七八成干时再摊铺第二层，厚度稍薄，照此摊放5~7层，待最表层达七八成干时，退火翻焙，即最上层和最下层翻转，使其均匀受热，干燥适度。由于受热条件使叶内多酚类化合物在热化作用下发生非酶性自动氧化，叶绿素遭到破坏，形成了黑茶色泽油黑且具有松烟香味的独特品质。

The roasting method is generally used for drying Dark Tea. Dark Tea is baked with strong firepower on the Seven Star Stoves. To spread the first-floor tea dhool when it reaches the proper temperature. Then respreads the second-floor tea dhool when tea leaves are 70%-80% dry. The thickness is slightly thinner. It should be spread over five to seven levels. When the top layer of tea is 70%-80% dry, turn it over to roast. Turn over the top and bottom floor of tea leaves to evenly heat and dry. The tea leaf's polyphenol compounds will react to the nonenzymatic automatic oxidation under thermalization to destroy the Chlorophyll. It will create the unique characteristics of dark-bloom dry tea and pine smoke for Dark Tea.

压 制
Compressing

黑茶可直接散饮，也可进行压制，是多种紧压茶的原料。压制是将初制好的毛茶通过加工、蒸压来对其塑形。由黑茶压制而成的砖茶、沱茶、饼茶、六堡茶等，深受我国部分民族地区人们的喜爱。

Dark Tea can be drunk directly and compressed. It is the law material for kinds of compressed tea. The step of compressing is to shape the law tea through processing and autoclaving. The ethnic minority areas in China love Brick tea, Tuo Tea (bowl-shaped compressed mass of tea leaves), Tea Cake, and Liu Bao Tea, compressed by Dark Tea material.

花茶
Scented Tea

　　花茶花香袭人，汤色明亮，叶底细嫩，最适宜清饮，也可加入适量蜂蜜，以保持其特有的清香。不同的花草配制成的茶营养成分不同，有不同的保健功效。对于平时久坐办公室，缺乏运动的上班族来说，花茶是具有天然的醒脑明目、提神保健功能的饮品。

　　The sweet scent of flowers drifts in the air. The tea soup of Scented Tea is fresh and bright. Those brewed leaves are fine and tender. It is suitable for light drinking or adding honey to maintain its unique fragrance. Its nutrient compositions are configured into tea with different nutritional components and different health care effects. They have different health functions. Scented Tea is the most natural drinking to refresh yourself, benefit your eyes, and relax the mind for sedentary office workers who lack exercise.

花茶的制作
Processing Techniques

原 料 Material

花茶是由精制后的茶坯和具有浓郁香气的鲜花窨制而成。茉莉花、玳玳花、玫瑰花、珠兰花、百合花、桂花等都可作为花茶的原料。质量上乘的花茶需要由当天采摘的成熟花朵制成。由于烘青茶的吸附力强，所以茶坯一般采用烘青绿茶，也有一些选用红茶和乌龙茶。

Scented Tea is processed with refined tea dhool and those flowers with strong perfume. The jasmine flower, citrus Aurantium, roses, orchids, lilies, osmanthus and so on can be used as Scented Tea materials. The highest grade Jasmine Tea material is the mature flowers plucked the same day. Due to the strong absorbability and hot air fixation, Green Tea is generally used as the tea dhool. Black Tea and Oolong Tea are also selected as tea dhool to make other teas.

配合不同鲜花制成的茶叶还有不同的保健功效：如茉莉花茶具减肥、润肠的作用；玫瑰花茶能调节气血、消除疲劳；菊花茶和金银花茶有清热解毒，疏风散热的效果，等等。

The different kinds of Scented Tea have different health healthy function. For example, Jasmine Tea has the effect of weight loss and moistening. Rose tea can regulate blood lipids and eliminate fatigue. Chrysanthemum and honeysuckle tea has the effect of heat-clearing and heat radiation.

窨制
Scenting

花茶窨制是将精制的茶坯与鲜花充分混合、静置，使茶叶充分吸收花的芳香的过程。茶叶表面有很多具有吸附力的空隙，气味清新，能与花香有效结合。茶坯与鲜花都要为其创造一定的外部条件才能达到最佳的吸香和吐香状态。茶坯含水量超过20%时，就基本失去了吸附能力；若含水量太低，则容易造成干燥。一般在含水量为5%时，其吸附能力最强。

The critical step of scenting is adequately mixing the tea dhool and flowers. Then set them for a while to help tea leaves fully absorb the fragrance of flowers. There are some spaces with absorbability on the tea surface Which has a fresh smell and combine flower scents effectively. Depending on certain external conditions, tea dhool, and flowers only can reach the optimum state to absorb the aroma and emanate incense. It will essentially lose absorbability if the moisture content of tea dhool surpasses 20%. It can become dry faster with lower moisture content. The absorbability is the strongest, with a 5% moisture content.

窨制前要对茶坯进行筛选和干燥，使其品质和湿度达到理想状态。鲜花吐香也需适宜的温度促进，所以在窨制期间，要适时翻拌茶堆，降低内部温度，使空气流通。直至花朵开始萎蔫，茶坯柔软，窨制才基本完成，还要注意筛去花渣。根据茶香的需要和成茶等级的不同还可以进行多次窨制。

Tea dhool need to be sifted and dried before scenting. It can help the quality and humidity degree to reach the ideal condition. The incense of flowers can only be emanated at a suitable temperature. The pile of tea leaves needs to be stirred and mixed at the right time during the scenting. It can help lower the internal temperature for better ventilation. Scenting is finished until the flowers begin to wilt, and tea dhool are soft. Then, pay attention to sifting the flour residue. The tea aroma degree request and different finished tea levels can be processed several times, scenting.

干燥和冷却
Drying and Cooling

在窨制过程中,茶坯不仅吸收了花的香气,同事也吸收了一定的水分,这就要求在窨制后对其进行干燥处理,防止霉变,利于储存。干燥后将花茶摊放,待其自然冷却,至此完成花茶的主要制作工序。

Tea dhool not only could absorb the flower's fragrance but also absorbs a certain amount of water. during the scenting process. Drying the tea leaves after scenting requires us to prevent going moldy. It is suitable for storage. After drying, we spread out the scent teas to cool down naturally. The primary production process of scent tea finishes after this step.

花茶的冲泡
The Brewing Step of Scented Tea

茶类篇
The Category of Tea

适用茶具
Appropriative Tea Ware

花茶种类不一，不同的花茶所选用的茶具也有不同的讲究。对于高档花茶，其品质特色和绿茶相似，茶叶在水中形态各异，袅娜多姿，所以可用无花透明玻璃杯冲饮，以便于欣赏其"茶舞"之翩跹。还可选用白瓷盖碗或带盖的瓷杯，以防止浓郁的花香散失。

There are many types of Scented Tea. Different teaware is suitable for kinds of Scented Tea. For high-quality Scented Tea has similar characteristics to Green Tea with various shaped leaves in the tea soup. They have a graceful gestures. So we can use transparent glasses to infuse Scented Tea to appreciate the dancing of tea leaves. You can also choose a white porcelain-covered bowl or a covered porcelain cup to infuse the Scented Tea to prevent the loss of strong and floral fragrance.

水 温
Water Temperature

　　花茶可用刚刚沸腾的开水来冲泡，水温在85℃～95℃为宜。水温偏低会影响花茶的香气和滋味，水温太高又会把茶中的"花"烫蔫。高档名优花茶的品饮虽以香气为重，但其外形也有很高的欣赏价值。透过玻璃杯欣赏干花在沸水中精美别致的翩翩飞舞之状也是不容错过的品茶乐趣。泡茶时，要先用温水将茶浸润一下，使茶汁更容易释放，然后再冲入沸水。

　　The Scented Tea can be brewed with just boiling water. And the suitable temperature is around 85℃～95℃. Lower temperature water could make weaken the fragrance and taste of Scented Tea. The flower wilting will appear with too high-temperature water. Even the aroma of the high-grade famous Scented Tea is more critical. It also has a high aesthetic value in the appearance of dry tea. You cannot miss appreciating the exquisite dance of the flower through a glass. It is joyful to enjoy tea. Warm and wet the tea before brewing it can help release tea juice and poured into the boiling water.

闷 泡
Brewing with the Lid

　　冲泡花茶时，注入沸水后一定要加盖，以免茶香散逸。热气集中在杯内，加速花香的释放，闷泡时间约3～5分钟，有的品种可以闷泡 5 ~ 8 分钟，让花茶更加出味。花茶的冲泡次数以 2 ~ 3 次为宜，一开茶饮后，留汤 1/3 时续加沸水，为之二开。如是饮三开，茶味已淡，香气流失，则不再续饮。

Put a lid on the top after pouring it into boiling water for brewing Scented Tea to avoid fragrance loss. The hot air concentrates inside the glass help accelerate the flower aroma release. The brewing time takes about three to five minutes. Some types can be brewed in five to eight minutes. It can help us to smell the fragrance of the flowers more quickly. The suitable brewing times for Scented Tea are twice to three times. Drink the Scented Tea until the water remains almost 1/3 inside the pot. They add the boiled water after the first drink. Then the brewing cycle continues three times until the tea tastes plain with a faint fragrance.

闻　香
Smelling

花茶吸附了鲜花的芬芳香气，以馥郁的花香为贵，品茶时重在闻香。闷泡过后，打开杯盖，随着热腾腾的水雾，浓烈的花香混和了茶香，立刻扑面而来，茶味与花香巧妙融合，相得益彰。这种香气纯正鲜活，如同给予杯中茶水灵动的精神，令人心旷神怡，未尝先醉。有兴趣者，还可凑着香气做深呼吸，充分领略愉悦香气，称为"鼻品"。茉莉花香被誉为花茶中春天的气息，有提神功效，可安定情绪、舒解郁闷。

The Scented Tea absorbed the flower fragrance. The particular type is rich fragrant. Focus on smelling while tasting tea soup. Open the cup lid after brewing for a while. The strong flower fragrance mixed with tea aroma is blowing with the hot water mist. Tea aroma cleverly integrated flower fragrance. They have an excellent complement to each other. This type of fragrance is pure and fresh. We will get drunk before tasting tea soup. If interested, you can breathe the fragrance deeply to appreciate the pleasant aroma fully. It is called Nose Tasting. The jasmine aroma is knownas the breath of spring among the Scented Teas. It can help refresh the mind and calm mood to relieve depression.

细 品
Tasting

　　花茶需要"品"。所谓的"品"其实很有含义，观其字形是由三个"口"组成，所以喝三口茶才是真正的品。品茶要在茶汤稍凉适口时，小口啜入，在口中稍事停留，以口吸气、鼻呼气相配合的动作，使茶汤在舌面上往返流动一两次，充分与味蕾接触，品尝茶味和汤中香气后再咽下，如是一两次，才能尝到名贵花茶的真香实味。此味令人神醉，正如古人所云"香于九畹芳兰气""草木英华信有神"。

　　好的花茶需要色、香、味俱全，通过三开茶汤的鼻闻、口尝和领略茶味的适口与否以及香气的生动，才能最终品饮出花茶的真味。

　　The Scented Tea needs to be tasted slowly to experience it fully. The structure of the Chinese character "Pin" is meaningful. It contains three rectangular frames that look like a mouth in Chinese. So there is a genuine taste of drinking tea soup three times. Tea soup needs to be tasted after being cooled with a better flavor. Sip tea soup and let the soup stay in your mouth. Tea soup flow back and forth on the tangue surfoce. Appreciate the tea fragrance and aroma of tea soup before swallowing. You can experience excellent quality Scented Tea two or three times a tasting procedure. The taste is intoxicated. the fragranc aromatic and orchid in Jiuwan and "the gross and trees are gods". The higher-grade Scented Tea is good in color, smell, and taste. We can fully appreciate the palatable and delicate tea fragrance by smelling, drinking, and experiencing it.

皇菊 Royal Chrysanthemum

"婺源皇菊"名称的来历传说可追溯至清朝年间。清光绪十六年（1891），原籍江西婺源的江人镜赴扬州任两淮盐运使，传说在他退休告老还乡之时，因政绩显著，光绪皇帝欲赏赐他千两黄金，被他婉言谢绝，只讨取皇家花园中作为药用的黄菊花带回婺源栽种，要效仿陶渊明"采菊东篱下"的悠闲生活。出人意料的是，因婺源的独特自然条件，使当年种植的黄菊花异常茂盛，而且浓香扑鼻，用水冲泡后入口甘甜，汤色金黄，韵味无穷，不但保存了其原有的药用功效，还可以直接当茶饮用。

The origin and legend of "Wuyuan County Royal Chrysanthemum " can be traced back to the Qing Dynasty. In the 16th year of Emperor Guangxu of the Qing Dynasty (1891), Jiang Renjing, originally from Wuyuan, Jiangxi Province, went to Yangzhou to serve as the salt transportation envoy of the two Huaihe rivers. It is said that when he retired and returned to his hometown, Emperor Guangxu wanted to reward him with gold, which he politely declined. He only asked for the yellow chrysanthemum used as medicine in the royal garden to be brought to Wuyuan County for planting. He wanted to follow Tao Yuanming's example and live leisurely under the eastern fence of picking chrysanthemums. Under the east hedge. Surprisingly, due to the unique natural conditions of Wuyuan, the yellow chrysanthemum planted that year is highly prosperous, with a strong fragrance, sweet entrance, golden soup color, and infinite charm. It can preserve the original medicinal effects, make tea, and drink it directly.

图片由茶人陈大华提供
Photos by courtesy of Chen Dahua

TEA PROCESSING

只有中国大地的富饶沃土，才能滋养出这么多异彩纷呈、美不胜收的茶叶吧，仅仅听它们的名字，便已是一场美好的检阅仪式。

China provides excellent fertile ground for nourishing wonderful and precious teas. It is so beautiful when we hear their tea names.

茶经三之造

Chapter 3: Tea Processing

茶造篇
TEA PROCESSING

采茶适宜在二月至四月进行。肥厚的芽叶,通常生长在含有碎石的土壤中,有数厘米长,状如刚刚抽芽的薇、蕨等植物,在清晨芽叶上还带着露水时采摘最好。略为瘦小的芽叶,多生长在草木丛中,有并发三枝、四枝、五枝的,要挑选叶片茂盛的采摘。采摘的时间要注意,下雨天不采,多云的天气也不采。只有晴朗无云时采摘才好,且当天便要将采摘的芽叶蒸、捣、烘烤、穿起、封存、晾干。

According to the circumstance, the suitable seasons for plucking tea leaves are February, March, and April. Those fat and thick tea leaves usually grow in the soil with gravel. Leaves are a few centimeters long and look like sprouting ferns. They are best picked early in the morning when the dew is still on the buds. The slightly thin and small-bud tea leaves grow in the grass and trees. There are some branches. Those leaves with flourishing glume can be selected and plucked. Its plucking and processing craft can be generalized as follows: avoiding picking in rainy days or the frost and cold dew weather. It can just be plucked on cloudless days. Those plucking tea leaves must be processed with steaming, tamping, baking, wearing, sealing, and drying on the some day.

茶叶产地
Tea Processing Area

山南地区，峡州产的茶最好，襄州、荆州的茶次之，衡州产的则差一些，金州、梁州的又更差一些。

Xiazhou County produced the best quality tea in the Shannan area. Xiangzhou and Jingzhou tied for second place. Tea processed in Hengzhou County is worse. However, Jinzhou and Liangzhou County produced the worst tea.

江南名茶
Famous Tea in the South of the Yangtze River

　　江南茶区是中国的四大一级茶区之一，年产量大概占全国总产量的三分之二。生产的茶类主要有绿茶、红茶、黑茶、花茶及各种特种名茶，比如西湖龙井、黄山毛峰、洞庭碧螺春、君山银针、庐山云雾等。"名山出名茶"，这句话在这里得到了集中的体现。但凡每一种名茶，无一不有历史，无一不有名胜，更无一不有文化背景和美丽传说。也许江南名茶就是因为江南的"千山、千水、千才子"而风流婉致。名茶通过名人的品饮而更加卓著，名人更在名茶之中淡泊明志。茶与人相得益彰，甚至可以说江南便是中国茶文化的滥觞。

　　The south of the Yangtze River tea area is one of the four largest first-class producing areas in China. Its annual yield accounts for about two-thirds of the national output. The main product types are green Tea, Black Tea, Dark Tea, Scented Tea, and some different varieties of famous special tea. Such as Xihu Dragon Well Tea, Yellow Mountain Fuzz Tip Tea, Dong

Ting Lake Green Spiral Tea, Jun Mountain Silver Needle Tea, and Lushan Mountain Mist Tea. Famous tea is cultivated infamous mountains. Each renowned tea has its favorite historical story, cultural background, and legends. That famous tea in the South of the Yangtze River is distinguished and admirable. They are more outstanding through drinking by celebrities. Celebrities show high ideals by simply living beyond utility by drinking those famous teas. Tea and people comple ment each other. We might even say the South of the Yangtze River is the original area of Chinese tea culture.

江南茶区
The South of the Yangtze River Tea Area

江南茶区位于长江中下游以南，石溪、大樟溪、梅江、连江以北，包括浙、湘、赣、苏和皖南、鄂南等地区，是目前中国绿茶生产最集中的茶区。

The South of the Yangtze River tea area is located in the middle and lower reaches of the Yangtze River and the north of Sixi, Dazhangxi, Meijinag, and Lianjiang County, including Zhejiang, Hunan, Jiangxi, Jiangsu, southern Anhui, southern Hubei, and other regions. It is the largest Green Tea production area in China.

地理特征
Geographical Features

江南茶区大多集中在低矮的丘陵地区，也有一些海拔较高的高山。土壤主要是红壤和黄壤，还有少量的冲积壤。

The South of the Yangtze River tea area is mainly in the low hilly area. However, some places are in the mountains with high altitudes. The soil underneath is primarily red and yellow soil with few alluvial soils.

茶区气候四季分明,全年平均气温约为15℃～18℃,冬季最低气温在-8℃左右,降水约有60%～80%集中在春季和夏季,秋季则较为干旱。该茶区是种植绿茶、红茶、黄茶和黑茶等茶类较为适宜的地域。

The tea area has four distinct seasons. The annual mean temperature is about 15℃～18℃. The minimum temperature in winter is around -8℃. About 60%～80% of rain occurs in the Spring and Summer. Autumn is a relatively dry season. The tea area is suitable for cultivating Green Tea, Black Tea, Yellow Tea, Dark Tea, and other tea kinds.

茶树品种
Tea Varieties

江南茶区的茶树以灌木型中叶种和灌木型小叶种为主，还包括少部分的小乔木型中叶种和小乔木型大叶种。其中小乔木型中叶种茶树的植株多为中等大小，树姿呈半展开状，分枝比较密集。

Those tea trees are mainly medium shrub leaf and small shrub leaf varieties, which grows south of the Yangtze River tea area. They also include a few small trees, medium leaf species, and small trees with large leaf species. Tea leaves are the medium size and semi-unfolded with dense branches. Among them, the tea of small tree type are medium in size, the posture is half spread, and the branches are relatively dense.

特产名茶
Specialty Famous Tea

江南茶区是中国绿茶产量最多的产区，其中有很多名茶都以其原产地命名。如产于浙江杭州西湖山区的西湖龙井，湖州市长兴县顾渚山的顾渚紫笋，以惠明寺一带为主要产区的惠明茶和余杭县径山的径山茶。江苏省有产于吴县洞庭山区的洞庭碧螺春，连云港市花果山的云雾茶，南京雨花台的雨花茶，无锡市的无锡毫茶。此外还有江西庐山的云雾茶，婺源县的婺源茗眉，以及湖南君山岛的君山银针，安化县的安化松针，高桥茶园的高桥银峰以及湖红功夫，

安徽黄山的黄山毛峰，祁门县的祁门红茶等。

The largest Green Tea production area in China is south of the Yangtze River tea area. Their city of origin names many famous teas. West Lake Dragon Well Tea grows in the West Lake mountain area in Zhejiang Province. Guzhu Mountain bamboo shoots tea grows in Guzhu Mountain Changxing County in Huzhou City. Huiming Green Tea is produced in the central area around the Huiming Temple and Jing Mountain Green Tea in Yuhang County. Dong Ting Green Spiral Tea is made in the Dong Ting mountain area in Wu County in Jiangsu Province. The mist tea grows in the Huaguo Mountain in Lianyungang. Rain Flower Tea grows in the Yuhuatai (The pronunciation of Yu Hua in Chinese is rain flower.) Nanjing city. Wuxi tip tea grows in Wuxi city. There are more types: Lushan Mist Tea, Wuyuan Brow Tea, Jun Mountain Silver Needle, Anhua Pine Needles Tea, Gaoqiao Silver Mountain Tea, Huhong Gongfu Tea, and Yellow Mountain Fuzz Tip, and Keemun Black Tea.

西湖龙井
Xihu Dragon Well Tea (West Lake Longjing Tea)

　　西湖龙井是中国十大名茶之一，属绿茶，因产于浙江省杭州市西湖龙井村周围群山，故此而得名。清乾隆皇帝游览杭州西湖时，盛赞西湖龙井茶，把狮峰山下胡公庙前的十八棵茶树封为"御茶"。一般爱喝绿茶的人，大都称道龙井茶，因其茶味醇厚，幽而不烈。西湖龙井茶是人、自然、文化三者的完美结合。

　　West Lake Longjing Green Tea is one of China's top ten famous teas in the green tea category. It is produced in the mountains around West Lake Longjing Village in Hangzhou City, Zhejiang Province, and is named after it. During the visit of Emperor Qianlong to West Lake in Hangzhou, he praised the West Lake Longjing Tea and named the eighteen tea trees in front of the Hu Gong Temple at the foot of Shifeng Mountain "Imperial Tea." People who love to drink green tea generally applaud the Longjing tea because it has a rich and mellow taste, which is light but not strong. West Lake Longjing Tea perfectly combines humans, nature, and culture.

饮誉世界的"国茶"
The National Tea Renowned in the World

　　西湖龙井茶的历史最早可追溯到唐代,当时著名茶圣陆羽所撰写的《茶经》中就有杭州天竺、灵隐两寺产茶的记载。北宋时期的龙井茶区已初具规模,南宋时期的杭州成了国都,茶叶生产得到了进一步发展。元代起,西湖龙井地区因风光幽静,且有甘泉香茶而广受文人雅士的推崇。到了明代,西湖龙井茶开始走出寺院,为普通百姓所饮用,此时的西湖龙井茶已是声名远扬,被列为中国名茶之一。清代的乾隆皇帝曾先后四次到龙井品茶赋诗,并将胡公庙前的18棵茶树封为"御茶"。此后,西湖龙井一直是清朝皇室的贡品。至民国初期,西湖龙井茶的种植已遍布西湖湖西、湖南各处,形成了"狮、龙、云、虎"四个主要的龙井产地。至此,西湖龙井茶成为中国名茶之首,驰名中外。

　　The history of West Lake Dragon Well tea can be traced back to the Tang Dynasty. Moreover, there is still evidence that West Lake Longjing Tea produced tea in history. It has been recorded in The Classics of Tea, edited by Lu Yu. The Xihu Dragon Well tea production area was shaped in the Northern Song Dynasty. Hangzhou city became a national capital in the Southern Song Dynasty. Tea production got developed at that time. Those refined scholars admired the quiet and secluded place of the Longjing area and sweet tea since the Yuan Dynasty. West Lake Longjing tea began to be spread outside the temple and drank by the ordinary people in the Ming

Dynasty. Xihu Dragon Well tea has become widely known and is listed as a famous Chinese tea. The emperor Qianlong went to the Longjing area four times to taste the tea and composed a poem. Then he gave the title of royalty tea tree to eighteen tea trees in front of Hugong Temple. Afterward, Xihu Dragon Well tea had always been the tribute of the royal family. Xihu Dragon Well Tea has already been cultivated in the west part and south parts of West Lake. It formed four major production areas of Dragon Well, named the lion, dragon, cloud, and tiger areas. Lion Peak Mountain, Longjing, Lingyin area, Five Clouds Mountain, Tiger Running Temple, and Meijia Dock area. West Lake Dragon Well tea has been the famous top tea of China. It is known across the world.

孕育名茶的环境
The Growing Environment of Famous Tea

　　茶区得天独厚的生态环境是培育名优茶品必不可少的条件。龙井茶区主要分布在杭州西湖西南侧的狮子峰、龙井、灵隐、五云山、虎跑、梅家坞一带。这里山峦叠翠，古树参天，四季分明，温度适宜，湿润多雾。茶区土壤为厚度适中、质地疏松、通透性好的微酸性砂质土壤，有机层深厚，养分充足，排水良好，施肥效果显著。茶树在这样优越的地势条件和良好的生态环境中可以持续平稳地生长，为充足的产量和优良的品质打下了良好的基础。

The unique and precious natural resource is essential for cultivating high-quality teas. The producing area is widely distributed in the Southwest of West Lake in Hangzhou cities, such as the Lion Mountain area, the Longjing area, the Ling Yin Temple area, the Five Clouds Mountain area, Tiger Run Mountain, and the Meijia Wu area. The mountain ranges rise and fall, with numerous peaks and cliffs with trees. Some lush ancient trees grow on the hill. It has four seasons with suitable temperatures and frequent humid, foggy weather. The thickness of the soil is moderate. The slightly acidic sandy soil is loose with good air permeability. It has a deep, nutrient-rich, and well-draining soil organic layer. The effect of fertilization is significant. Tea trees can keep growing steadily with these right conditions. And it lays a good foundation for sufficient output and good quality.

图片由茶人戚英杰提供
Photos by courtesy of Qi Yingjie

独特的工艺
Unique Processing Techniques

绿茶的制作一般都要经过采、晾、揉、炒等数道工序，龙井茶外形的"扁、平、光、滑"以及"色、香、味"等独特的品质，就得益于精湛的炒制技术和独特的加工工艺。

Green Tea has to undergo some phases, such as plucking, drying, rolling, and roasting. At the same time, however, the feature leaves are flat and smooth, with a beautiful shape and unique quality in color. Exquisite roasting technology and special processing techniques create its smell and taste.

极品龙井的炒制
Processing Method of Super Grade Longjing

龙井茶的制作流程与绿茶基本相同，形成其优异品质的关键就在于其复杂精湛、独具特色的炒制技术。由于机械化炒制的技术不过关，炒制出来的龙井茶外形粗糙，内质不佳，失去了传统茶的醇厚风味，只能作为中低档龙井茶。因此，为了保证成茶的品质，特等和上等龙井仍采用传统手工炒制的方法，且级别越高，锅温越低，投叶量越少，炒制的手法也越轻。

Its production processing is the same as Green Tea. The sophisticated and unique stirring technique is critical to forming its excellent qualities. Due to the mechanized stirring, the Longjing tea is coarse with worse quality. It loses traditional tea's mellow taste and can be below medium grade. Longjing Tea still needs to be processed with the traditional handmade stirring to maintain tea quality, super grade, and first-class. The temperature should be lower for higher-grade tea. Fewer tea amounts should be with less power.

 天地精华
The Essences of Heaven and Earth

黄山毛峰，又名黄山云雾茶，属绿茶烘青类，是中国十大名茶之一。该茶外形微卷，仿若雀舌，绿中带黄，银毫毕显，带有金黄色鱼叶，因此也被称为"黄金片"。黄山毛峰的产区在安徽黄山风景区和相邻的汤口、充川、芳村、岗村、扬村、长潭一带。因为茶叶身披白毫，芽尖锋芒，来源于黄山高峰，所以被命名为黄山毛峰。

Yellow Mountain Fuzz Tip Tea is also called Yellow Mountain Mist Tea. It belongs to the category of hot air fixation Green Tea. It is one of the ten most famous types of tea in China. The dry tea is slightly curly and shaped like a bird's tongue with plenty of silver fuzzes. The color is green with yellow. It is called golden leaf for its some fish leaves. The production area for Yellow Mountain Fuzz Tip tea is mainly located in the beautiful scenic city of Huangshan Mountain (Yellow Mountain) in Anhui Province and some neighboring Tangkou, Chongchuan Fang village areas, Gang village, Yang village, and Changtan County. Because of the origin place, the white fuzz is covered with tea leaves, and the sharp bud tip is called Yellow Mountain Fuzz Tip because of the origin place.

黄山为中国东部的最高山峰，素以奇、险、深、幽而闻名于世。黄山毛峰茶园就分布在海拔1000米左右的半山周围，或分布在坡度达30°～50°的高山深谷中。那里气候温和，雨量充沛，空气湿润，日照时间短，土壤肥沃且呈酸性，质地疏松，具有良好的透水性，磷钾和有机质含量也十分丰富，适宜茶树生长。正是这种优越的生态环境，为黄山毛峰自然品质的形成创造了极其良好的条件。

The Yellow Mountain is the highest in the east of China. It is famous for its strange, risky, deep, and secluded world. The tea gardens of Yellow Mountain Fuzz Tip are distributed halfway up the mountain, around 1000 meters in elevation. Or some are distributed in deep valleys with a slope of 3050 meters. The district has a mild climate, plentiful rainfall, moist air, and short sunshine. The fertile soil is acidic and loose with good perviousness. The content of available P and available K is rich. Therefore, it is suitable for tea growth. This ideal environment creates conditions for the excellent natural quality of Yellow Mountain Fuzz Tip tea.

杯中景象万千
Pictures in the Tea Cup

  取茶 3 ~ 5 克，以 80℃ ~ 90℃的水温冲泡，用玻璃杯或者白瓷茶杯皆可。先投茶，然后注入 1/3 杯水，待 3 分钟左右，茶叶舒展之后再将水加足。一般可续水冲泡 4 ~ 6 次。品质佳的毛峰茶在冲泡后，雾气凝顶，芽叶竖直悬浮于汤中，之后徐徐下沉，芽挺叶嫩，景象万千，茶汤清澈，叶底明亮，嫩匀成朵。更有趣的是，用黄山泉水冲泡黄山毛峰茶，即使茶汤经过一夜，第二天也不会在茶杯中留下痕迹。

  Take thirty-five grams of dry tea leaves to brew with eighty to ninety degrees water. The glasses cup or white porcelain cup can also be brewed tea. Throw tea leaves, pour them into third water, and then wait for about three minutes. Add adequate water while the tea leaf is about to unfold. It can usually be brewed four to six times. Water vapor condenses to liquid water on the top, and a few bud tea leaf is suspended in water, then sink slowly. Tea leaf is straight, bright and tender and presents different pictures. The tea soup is clear and with bright brewed tea leaves. Interestingly, the tea soup brewing Yellow Mountain Fuzz Tip tea with Yellow Mountain spring water can stay overnight. It won't leave a trace inside the teacup.

皇帝赐名
Name-Granting by Emperor

碧螺春又称"吓煞人香",产于水汽升腾、雾气悠悠的江苏省吴县太湖的洞庭山碧螺峰,是中国十大名茶之一。

The Dong Ting Green Spiral Tea is also called Xia Sha Ren Xiang, which means its fragrance is too strong. It is cultivated in Dong Ting where water vapor rises and fog is long. Mountain, draped in purple heather, and full of the damp haze in the Taihu lake region in Wu County, Jiangsu Province. It is one of the top ten famous teas in China.

很多品饮过碧螺春的人,都会为它的嫩绿隐翠、清香幽雅和绝妙韵味所倾倒,但很少有人知道其名称的由来。在《苏州府志》中有这样一段话:"洞庭东山碧螺石壁,产野茶几株,每岁土人持筐采归,未见其异。康熙某年,按候采者,如故,而叶较多,因置怀中,茶得体温,异香突发。采茶者争呼:吓煞人香!茶遂以此得名。"

People who drink Dong Ting Green Spiral Tea will be charmed by its jade green color, fragrance aroma, and fantastic Yun flavor. But few people know the origin of the name. There is a quote written in a book called Suzhou Provincial Gazetteer. The cliff, named Bi Luo in the East Mountain area of Dong Ting, produced several wild tea strains. Those local farmers holding baskets didn't find it different when they plucked every day. A farmer usually picks up tea leaves during the Qing Emperor Kangxi years. He put some leaves in front of his chest. Tea leaves send out their perfume and sweet aroma while it is warm. Tea farmers shouted,Scary and fragrant: Xia Sha Ren Xiang and then it is named after this.

后人评价说:"此乃康熙帝取其色泽碧绿,卷曲似螺,春时采制,又得自洞庭碧螺峰等特点,钦赐其美名。"由此可见碧螺春深受康熙皇帝的喜爱,成为御用的贡茶。

Later people commented: "This is the color of Emperor kangxi, which is green, Curled like a snail, picked in spring, and obtained from the characteristics of Dongting Biluo Mountain, which gave him a good name."

一嫩三鲜
Tender and Fresh

碧螺春的品质特点是：条索紧细重实，似螺旋形卷曲，茸毛披覆，香气浓郁，滋味甘醇，汤色清澈碧绿，叶底嫩绿明亮，素有"一嫩三鲜"之称。当地茶农将碧螺春生动地描述为"铜丝条，蜜蜂腿，香果味，浑身毛"。

The quality characteristics of Green Spiral Tea are tight and heavy, curly spiral shape, covered with fuzz, strong aroma, mellow taste, clear Green Tea soup, and bright and tender brew leaf. It has long been known as one point tender with three parts fresh. Local tea farmers gave a graphic description of Green Spiral Tea. It looks like copper wires and bee legs with fruit fragrance and full of hair.

特殊的品饮方法
Special Drinking Method

品饮碧螺春宜采用质地细腻的白瓷杯或透明纯净的玻璃杯，先放入70℃~80℃的温开水，然后取少量茶叶投入水中，顿时出现"雪浪喷珠"的场面。其后，芽叶全部沉入杯底，杯底一片碧绿，好似"春染海底"，但此时茶汤尚无茶味，只有将水倒掉2/3时，才闻茶香袭人，这时再冲入滚水，茶叶则完全展开，渐渐舒展成一芽一叶，水色淡绿如玉，呈现"绿满晶宫"的景象。

图片由茶人施跃文提供
Photos by courtesy of Shi Yuewen

Using a fine white porcelain cup or a pure glasses cup to drink Green Spiral Tea is appropriate. Add the 70℃ ~ 80℃ water into the cup before putting a few tea leaves. Tea leaves instantly fall into the bottom and turn over in the cup. And the cup bottom is a rich green. But tea soup is plain and thin at this time. It has a strong fragrance aroma until you pour two-thirds of the water. Then add boiled water into the cup. Tea leaves will start to unfold to be one bud with one leaf. The tea soup color is light green as polished jade, showing us a green palace vision.

此时，碧螺春的色、香、味、形俱达到最佳状态，茶汤清冽，茶香清新，味道甘爽。先观其形，而后细品之，可以发现头酌汤色清淡，味幽香鲜雅；二酌汤色翠绿，味道芬芳醇美；再酌汤色碧清，香郁回甘。

The color, aroma, taste, and appearance will reach their optimal state at this time. The tea soup is clear, and the tea aroma is fresh and sweet. We can appreciate its appearance and taste tea soup next. We can find that the first sip is plain with a clear and fresh aroma. The second sip is jade green with a mellow and fragrance taste. The tea soup is green and clear for the next sip and has a sweet after taste.

祁门红茶
Keemun Black Tea

由绿转红
From Green to Red

祁门功夫红茶也被誉为"王子茶",又简称"祁红",产于中国安徽省西南部黄山支脉的祁门县一带,素以香高形秀而享誉海内外。1875年,祁门红茶创制,以功夫红茶为主。

Keemun Black Tea is also named Prince Tea and Keemun Black for short. It originated from the Qimen County offshoot of the Yellow Mountain range southwest of Anhui Province, China. Keemun Black Tea is internationally best known for its high aroma and beautiful shape. In 1875, Keemun Black Tea was created. Keemun Black Tea is one of the most outstanding teas among Gongfu Black Teas.

祁门茶叶早在唐代就已出名。据说，在清代光绪前，这里并不生产红茶，只盛产绿茶。1875年，黟县人余干臣从福建罢官，来到德县尧渡街设立茶庄，模仿"闽红"制法试制红茶。1876年，他再次来到祁门扩大生产和收购，使祁门一带逐渐改制红茶，并大获成功。由于祁红的价格高、销路好，人们纷纷响应改制，逐渐形成了祁门红茶的规模，距今已有100多年的历史。

Keemun Black Tea has already been famous in the Tang Dynasty. It is said that no Black Tea was produced. Just Green Tea was processed before the region of Emperor Guang Xu. In 1875, Governor Yu Ganchen from Yi County was dismissed from office in Fujian Province and then came to De County to set up a tea shop on Yaodu Street in De County. He imitated the processing technique of Fujian Black Tea to try to create a new style of Black Tea. He came back to Keemun County to expand production and purchasing. Then farmers in the Keemun area started to produce Black Tea. It was a big success. Due to the higher price and good sales of Keemun Black Tea, people followed suit to make Black Tea. It gradually formed a scale with broad marketing space, developing prospects, and a history of more than one hundred years.

扬名天下
Recognized around the World

祁门红茶是中国传统功夫茶之一，其条索紧细秀长，汤色红艳明亮，特别是其香气清新芬芳，馥郁持久，似蜜糖香，隐伏果香，又蕴藏有兰花香，口感醇厚，汤中带香，香中伴甜，回味隽永。祁门红茶自1875年问世后不久就享誉国际市场，成为中国传统的出口珍品，销往东南亚、北欧、英国、德国、美国和加拿大等50多个国家和地区。

Keemun Black Tea is one kind of the Chinese traditional Gongfu Tea. Its tea leaf is tight and slender. And the tea soup is red brilliant and clear. Especially its clear and fresh fragrance is high and sharp. It has a honey-sugar and fruit flavor and a unique orchid fragrance. The taste of tea soup is mellow and thick with the scent. It is a delightful drink. It has been famous soon in the international market since it was first introduced in 1875 and has become a Chinese traditional export treasure. It has been sold to more than 50 countries and regions, such as Southeast Asia, Northern Europe, England, Germany, and America, Canada.

仅有100多年历史的祁门红茶，在全球种类众多的红茶中已然独树一帜，长盛不衰，以其"形美、色艳、香高、味醇"四绝在国际市场上占有重要的地位。在国际茶人的认同和推崇下，中国的祁门红茶与印度的大吉岭茶和斯里兰卡乌伐高地的季节茶并列为世界

上最出众的三大高香茶。

Even though Keemun Black Tea has only one hundred years of history, it has a unique and enduring style, distinguished for progressive development and growth among the world's several Black Teas. Its four unique points, appearance, color, aroma, and taste, are famous worldwide. With the high identity of international tea masters, Chinese Keemun Black Tea, Indian Darjeeling Black Tea, and Sri Lanka Seasonal Ceylon Tea have become the three most outstanding Black Teas globally.

华南名茶
Famous Tea in South China

华南茶区是中国的四大一级茶区之一，由于其独特的气候条件，也是中国最适宜茶树生长的地区。其得天独厚的地理条件，使得华南茶区名茶辈出，如安溪铁观音、武夷大红袍、福建大白毫及产于台湾的冻顶乌龙茶，等等。华南名茶以乌龙茶、红茶、花茶和白茶为主，其中乌龙茶和红茶的品饮方法更是独具一格，开创了中国功夫茶的品饮方法。相对江南名茶的贵族气息和频繁地出入宫廷，华南名茶与平民百姓和具体的民生走得更加贴近，在阐述中国茶文化的生活方面上乃至于中国人的生活方式上，都有着不可代替的地位。

South China tea cultivation area is one of the first four-grade tea areas. With its unique climatic conditions and the most suitable environment condition for cultivating tea trees, this area could produce many types of famous teas, such as Anxi Tie Guan Yin Tea, Wuyi Mountain Da Hong Pao Cliff Tea, Fujian Province White Tea, and Taiwan Province Dong Ding Oolong Tea, and so on. Famous teas are mainly Oolong Tea, Black

Tea, Scented Tea, and White Tea in South China. The unique brewing and drinking method for Oolong Tea and Black Tea is unique for enjoying Chinese Gongfu Tea. Compared to that famous tea with an aristocratic feature in the South of the Yangtze River, South China tea is close to people's life. Therefore, it occupied an irreplaceable position in explaining Chinese tea culture.

区域范围
PLANT AREA COVERAGE

华南茶区包括福建雁石溪、大漳溪，广东连江、梅江，广西浔江、红水河，云南南盘江、保山、无量山、盈江以南区域，其中包括福建东南部、广东中南部、广西壮族自治区、海南省、云南南部、湖南南部及台湾等地区。这些地区均属亚热带及热带气候，大部分地区高温多雨且土壤肥沃，是我国较为适宜茶树生长的地区。

South China tea cultivation area in the Yanshi Creek area and Dazhang Creek area in Fujian Province, Lianjiang River area and Meijing River area in the Guangdong Province, Xunjinag River, Hongshui River, and Yunnan Province Panjang River area, Baoshan Mountain, Wuliang Mountain, south area of Yingjiang River. It includes the southeast part of Fujian Province, the middle south part of Guangdong Province, Guangxi Zhuang Autonomous Region, Hainan Province, the southern part of Yunnan Province, the southern part of Hunan Province, and Taiwan Island. These regions have a subtropical climate and tropical climate. Most area has a high-temperature environment and fertile soil. Therefore, it is the most suitable growth area for tea in China.

地理特征
Regional Geographical Features

华南茶区南部属热带季风气候，最主要的特点是高温多雨，长夏无冬。年平均气温为19℃～22℃，最低月份（1月）平均气温为7℃～14℃。年降水量大约是1200～2000毫米，为中国茶区降水量之最。茶树年生长期达10个月以上。

The southern part of the South China tea cultivation area is tropical monsoon climate. Its main characteristic is both high temperature and usually heavy rainfall. The summer will last for a long time. On the other hand, it never gets too cold. The yearly average temperature is about 19℃～22℃. The lowest monthly temperature is 7℃～14℃. The annual precipitation is about 1200～2000 millimeters. It has the most significant yearly rainfall among China tea cultivation areas. Its growth cycle is as long as more than ten months.

茶区北部属亚热带季风气候，最主要的特点是温暖湿润。全年只有春、夏、秋三个季节：春季多雨；夏季热而长，多台风暴雨；秋季雨水较少，较为干燥。年均降水量在1500毫米以上，主要降水集中在5～10月，约占年降水量的70%～80%。

The northern part of the tea area is subtropical monsoon climate. Its main characteristic is warmth in humid conditions. It only has three seasons in a year wet spring, hot, and long-term summer with plenty of storm rain and dry autumn. Between May and October, the annual precipitation is above 1500 millimeters, which makes up approximately 70% ～ 80% of yearly rainfall.

茶区土壤除了砖红壤外，部分地区还分布有红壤和黄壤，在森林植被的覆盖下，土壤肥沃，有机质含量丰富。但是，如果植被遭到破坏，土壤暴露，就很容易受到雨水浸溶，有机质分解很快，含量也会迅速降低。因此，在开辟茶园时，要合理规划和使用已经开垦的土地。

Besides the granitic latosol, red earth and yellow earth are in parts of the cultivation area. The soil is fertile with the abundant herbaceous organic matter under the natural vegetation cover. But if the eco-environment is destroyed to expose good soil, it is easier to be awash by rain. Moreover, the organic matter can be quickly decomposed, and its content could reduce soon. There fore it would be best to make proper plans to cultivate the newly reclaimed land before opening a tea plantation garden.

茶树品种
Tea Varieties

华南茶区茶树品种资源比较丰富，主要有乔木型和小乔木型的大叶种，少数地区也有灌木型小叶种的分布。乔木茶对环境要求很高，需要在没有污染、天然纯净的自然环境中才能孕育出品质优良的大叶种乔木茶。这类品种植株十分高大，主干分明且粗壮，分枝部位高，叶片大，结实率低，所以茶叶的采摘比较困难，价格也比较昂贵。

The southern part of the South China tea cultivation area has abundant resources. There are medium arbor leaf and arbor big leaf species. And there are also some shrubby middle-small leaf varieties in a few other areas. It has stringent requirements on the environment. High-quality tree tea only can be well cultivated in unspoiled, natural, and purely natural environments. The plant is very tall, with some strong trunks. The branch location is high. It has a big leaf size with a low seed-set rate. So those fresh tea leaves are hard to pluck. The price is also more expensive.

特产名茶
Specialty Famous Tea

华南茶区因其适宜的气候环境和肥沃的土壤条件,茶叶产区分布广泛,盛产的名茶也是不胜枚举。如产于广东潮州的凤凰单枞、英德市的英德红茶、仁化县的仁化银毫茶,福建省安溪的铁观音、武夷山的大红袍、福鼎的贡眉、永春县的永春佛手,广西凌云县的凌云白毫、苍梧县六堡山区的六堡茶、桂林的毛尖等。其中很多都是长期远销海外的优质名茶。

With the feasible climate, fertile soil condition, and widely distributed teas of the southern part of the South China tea cultivation area, famous types are too numerous to mention, such as Fenghuang Dancong Oolong Tea from Chaozhou City, Guangdong Province, Yingde Black Tea from Yingde City, Renhua Silver White Tea from Renhua County, Tie Guan Yin Oolong Tea from Anxi County in Fujian Province, Wuyi Mountain Da Hong Pao Oolong cliff tea, Gongmei Tea from Fuding County in Fujian Province, Yongchun Foushou Tea from Yongchun County, Lingyun White Tea from Lingyun County in Guangxi Province, Liu Bao Dark Tea from Liu Bao area in Cangwu County and Guilin City Tip Tea and so on. Some famous high-quality teas have been exported overseas for a long time.

台湾名茶
Taiwan Province Famous Tea

台湾几乎全省都产茶。中南部有南投渔池和埔里茶区的日月潭红茶、鹿谷的冻顶乌龙、名间的松柏常青茶、嘉义的阿里山珠露茶；东部有台东鹿野的太峰高山茶和花莲的天鹤茶；北部的台北县出产文山包种、木栅观音、三峡龙井、桃竹苗茶区的桃源县龙泉茶、新竹县东方美人；还有海拔较高的高山茶园的高山茶……这些名优茶品以其独特的风味和醇厚的口感得到海内外广大茶叶爱好者的喜爱。

Tea is cultivated almost all over the Taiwan Province area. There are Riyuetan Pool Black Tea from Nantou County Fish Pond and Puli Tea Area in south-central Taiwan, Dong Ding Oolong Tea from Lugu County, Pine Tea from Mingjian Township, Pearls dew tea from Ali Mountain in Jiayi Region, in the east, there are Taifeng Mountain Oolong Tea in Luye County in southeastern Taitung County, and Crane Tea from Hualian County. In addition, northern Taibei County produced Wenshan light Oolong (Pouchong), Mushan Tie Guan Yin, Three Gorges Longjing Tea, Longquan tea from Taoyuan County, in Taozhu Miao district and Oriental Beauty Tea from Xinzhu County, and some Alpine tea (High Mountain Oolong Tea) from Alpine Tea Gardens. With their unique and mellow taste, these famous teas are appreciated by tea lovers worldwide.

安溪铁观音
Anxi Tie Guan Yin Tea

主要产地
Main Producing Areas

安溪，是中国乌龙茶的故乡，也是世界名茶铁观音的发源地。安溪凭借着其悠久的历史，丰富的资源，众多的品种，特有的制茶工艺和丰富的制茶经验，成为中国茶叶宝库中一颗耀眼的明珠。

Anxi County is the hometown of Chinese Oolong Tea. It is also the home of the world-famous Tie Guan Yin Oolong Tea. With its long history, rich resources, varieties of types, unique processing techniques, and rich processing experience, Tie Guan Yin Oolong Tea has been a sparkling pearl among Chinese teas.

名扬四海
Well-Known in the World

铁观音茶又称闽南乌龙,系乌龙茶中的珍品,兼有红绿茶的特点,原产于福建省南部安溪县西坪镇。铁观音出现于清雍正四年(1725)前后,后被传播到我国台湾、港澳地区,以及越南、泰国、印尼、新加坡等国家。

Tie Guan Yin Oolong Tea is also called Southern Fujian Oolong Tea. It is a treasure tea with the characyeristies of in Oolong Tea. It also has excellent features for Black Tea and Green Tea. It is native to the town of Xiping in Anxi County in the south of the Fujian Province. It is one of the Chinese national teas and world-famous tea. Tie Guan Yin Oolong Tea was created in Anxi around the fourth year of Emperor Yongzheng (1725). Then it spread to Taiwan, Hong Kong, Macao, Vietnam, Thailand, Indonesia, Singapore, and other countries and regions.

七泡有余香
Aromatic Smell after Seven Times Brewing

优质铁观音条索卷曲紧实,呈颗粒球状,色泽鲜润,叶表带白霜。其茶汤色泽金黄,浓艳清澈,叶底肥厚,具有丝绸般的光泽。茶汤醇厚甘鲜,入口甘甜略带蜜香,香气浓郁持久。

The high-quality Tie Guan Yin tea leaf is curly, tight, and heavy in pellet shape. Its color is blooming. Tea leaf is covered with white frost. The tea soup is golden bright, heavy, and clear. Brewed leaves are fat and thick with shiny silk hair. Tea has a mellow, sweet, and honey aroma. It has a strong and lasting fragrance.

近年来，经实验证明，乌龙茶中以安溪铁观音所含的香气成分种类为最多，而且中、低沸点的香气成分所占比重大于其他品种的乌龙茶。因此，安溪铁观音以其独特的香气令人心醉神往，享有"七泡有余香"的美誉。

In recent years, Anxi Tie Guan Yin Oolong Tea is one of tea types which has the most aromatic components among teas, based on the investigation research conducted in China and other countries. And its proportion of aromatic components with lower boiling points is much more significant than the different kinds of Oolong Tea. Accordingly, Anxi Tie Guan Yin tea charmed everyone with its fragrance and has been long reputed as the lingering odor feature after seven times brewing.

铁观音的"音韵"
IN-flavor of Tie Guan Yin

铁观音的品质特色除外形特征以外，尽可以用具有"音韵"来概括。"音韵"的全称是"观音韵"，无此不成铁观音。铁观音冲泡后，香气扑鼻，汤色同绿豆水，滋味鲜美，令人回味，而"音韵"

就是来自铁观音特殊的香气和滋味。铁观音的香气馥郁清高，鲜灵清爽，犹如空谷幽兰，滋润心脾，令人兴致盎然；铁观音入口醇厚鲜香，顺喉咙滑下，清爽甘甜，余味无穷，令人烦恼顿失。有人说，品饮铁观音中的极品——观音王，有超凡入圣之感，仿佛羽化成仙，将一切俗事抛于脑后，这至真至妙的感受恐怕就是人们将铁观音独特的风韵命名为"观音韵"的来源。

All unique quality characteristics of Tie Guan Yin Tea can be described as IN-flavor, except the shape characteristics. The IN-flavor can also be described as the Guanyin flavor. All Tie Guan Yin tea has this

unique flavor. A sweet smell assailed the nostrils when brewed Tie Guan Yin Oolong Tea. Tea soup's color is similar to mung bean water. The taste is fresh, delicious, and memorable. IN-flavor is created from its unique aroma and superior taste of Tie Guan Yin. Tie Guan Yin has a clear and high fragrance with mellow, thick, and sweet aftertaste. Worries can go away with the aroma. Some people say drinking the highest grade Tie Guan Yin King tea would obtain the sublimation and the eternal. All worldly affairs can be left behind. That is the reason that we named it Tie Guan Yin King Tea.

武夷岩茶
Wuyi Rock Tea

高人精心制作
Made by Master

　　武夷岩茶的制作可追溯到汉代，经过历代的发展沿革而成，至清代达到鼎盛时期。武夷岩茶是由武夷山独特的生态环境、气候条件和精湛的传统制作技艺造就的，其传统制作流程共有晾青、做青、杀青、揉捻、烘干、毛茶、归堆、定级、筛号茶取料、拣剔、筛号茶拼配、干燥、摊凉、匀堆、装箱等十几道工序，环环相扣，缺一不可，其细致繁复为武夷岩茶所独有。

　　The traditional production of Wuyi Rock Oolong Tea can be traced back to the Han Dynasty. It has been completed with the development of dynasties. The processing craft of Wuyi Mountain Oolong Tea experienced its peak development in the Qing Dynasty. Wuyi Mountain Rock Oolong Tea is made by its unique ecological environment and climatic conditions of the Wuyi Mountain area, with exquisite traditional craftsmanship. Its traditional processing technique can be divided into fresh leaf airing, fixation, rolling, drying, primary tea process, bunching, grading, reclaiming the material, extracting, blending, roasting, cooling, uniform-blending and packing. Many various steps could be linked together. Not a single one of these steps can be dispensed. This complete and meticulous processing is unique to Wuyi Rock Oolong Tea.

武夷岩茶的制作方法，汲取了绿茶和红茶制作工艺的精华，加上特殊的技术处理，使其独特的岩韵更加醇厚突显。这是武夷山历代茶农的智慧结晶，有着丰富的实践经验和独特而高超的技艺。2006年6月，武夷岩茶的制作工艺被列为首批"国家级非物质文化遗产"。

Wuyi Rock Oolong Tea's processing has absorbed the essence of Green and Black Tea processing techniques. With our special treatment, its unique YEN flavor is highlighted. It is the tea farmers' intelligence treasure in the Wuyi Mountain area of all time. It has a rich processing experience and a unique and masterful technique. Wuyi Rock Tea Processing Technique was listed in the National Intangible Cultural Heritage list in June 2006.

"还阳"萎凋
Withering

萎凋是形成岩茶香味的基础，可分为日光萎凋和加温萎凋两种。萎凋原则上是"宁轻勿过"，操作手法要轻，不能损伤梗叶，这样才能有利于恢复一部分弹性，俗称"还阳"。

The aroma of rock tea is created during the withering. It can be divided into two types: daylight withering and heating withering. The principle of withering should be followed as below. It would be relatively lightly than over withered. Tea leaves should be handled lightly without damage. It can help us to keep leaves to restore their elasticity. It is called revive.

做　青
Fine Manipulation

做青是武夷岩茶的制作过程中特有的工序，也是形成"三红七绿"的重要环节。该工序具有费时长、要求高、操作细、变化多等特性。做青的方法没有完全固定的模式，而是青变即变，气候变即变，需要变则变，根据品种、萎凋程度和当时的气温、湿度，以及后续工序的要求而采取不同的处理手段，俗称"看青做青"，这样才会使岩茶具有独特的风格和品质。

Making green is a specific step in the process of Wuyi Mountain Rock Tea. It is also vital to create and form the feature of Three Red and

Seven Green. This step provides some excellent features for a long time consuming, high requirement, and meticulous operation and diversity. There is no specific processing instruction for making green. Wuyi Mountain Rock Tea is made with color and weather changes. Tea leaves can be handled according to variety, withering degree, temperature, humidity, and requirement of the follow-up processes in different ways. It is commonly known that making green should be held with the color change. Rock Oolong Tea will have unique characteristics and extraordinary quality with these stops.

武夷岩茶的外形条索壮结、匀整，色泽绿褐鲜润，冲泡后茶汤呈深橙黄色，清澈艳丽；叶底软亮，叶缘朱红，叶心淡绿带黄，滋味兼有绿茶的清香和红茶的甘醇。茶性温和，不燥不寒，久藏不坏，香久益清，味久益醇。泡饮时宜选用小壶小杯，小啜细品。因其香味浓郁，冲泡多次后余韵犹存。

Wuyi Rock Oolong Tea leaf is bold, evenly, and greenish-brown. The tea soup is deep orange-yellow, clear, and brilliant. Brewed tea leaves are soft and bright. The leaf edge and body of the brewed tea are bright red. The leaf center is light green and yellow. It has a clean aroma of Green Tea and the sweetness of black tea. The character is mild, with no irritation. It can be kept for a long time. The fragrance will be clear with a mellower taste. Using small pots and cups to brew tea and drink tea soup is suitable. Sip tea soup and smell the fragrance slowly. It is a very unforgettable tea with its strong aroma.

品种繁多
Wide Varieties

武夷岩茶的品种繁多，特征各异。对其品种辨别主要就是从茶树的枝、干、叶，以及成品茶的外形、香型、茶汤、味道等方面进行。各品种间的差异有的较为明显，但大多茶叶区别细微，不易辨别，只有经过长期的观察、品饮和比较才能分辨。目前武夷岩茶的主要品种有大红袍、肉桂、水仙、白鸡冠和奇种等。

There are varieties of Chinese Wuyi Mountain Rock Tea with different features. It is mainly from the stem, trunk, leaf and appearance, aroma type, tea soup, and taste of finished tea to identify varieties. Some noticeable differences are between varieties. You cannot easily distinguish the fine distinctions for most Wuyi Rock Tea. The difference only can be found with a long time of observation, drinking, and comparison. The main varieties of Wuyi rock tea are Wuyi Mountain Da Hong Pao Cliff Tea, Rou Gui Tea, Shui Xian Tea, Bai Ji Guan Tea, Qi Zhong Tea, and so on.

水 仙
Shui Xian Tea

水仙属无性系小乔木型，大叶类，晚生种。其株高大直立，叶大者发芽早，叶长者发芽迟，叶面平滑且略带有绿色油光，边缘锯齿较深。花期较早，花朵多而大，红白色，不易结实。

Shui Xian Oolong Tea is a tea clone of arbor form with big leaf and late-sprouting species. Its plant is tall and erects with a big leaf. Those big size leaf plants only take a few more days than the aged leaf species of Shui Xian to sprout. The leaf is smooth and slightly green, and glossy. It also has an earlier inflorescence. Those numerous flowers, red and white, are big. This species is hard to bear fruit.

成品茶特征：条索肥壮，色泽乌绿油润，部分叶背常现沙粒，叶基宽扁，香似兰花，汤色浓艳呈金黄色，滋味醇厚，回味甘爽，叶底软亮，朱砂红边明显，较耐冲泡。

The characteristics of Shui Xian tea are as follows: fat and bold leaf, black-green, and bloom; the underneath has the form of frog skin and powdery white spots. Phyllopodium leaf is broad and wide. It has a natural fragrant orchid sweetness. The tea soup is bright golden with a mellow taste. It is a sweet aftertaste. Those brewed tea leaves are soft and bright. The cinnabar color red edge is apparent. You can brew this tea several times.

肉　桂
Rou Gui Tea

　　肉桂是武夷岩茶的当家品种之一，原为武夷名枞之一，属无性系灌木型。茶树高约1.6米，且冠大、干粗、枝叶繁密。叶片光滑椭圆，厚而脆，呈浓绿色，叶尖钝，叶缘内翻成瓦筒状。萌芽力很强，花朵多且小。

　　Rou Gui Oolong Tea is one of the leading varieties of Wuyi Mountain Oolong Tea. It is one of the famous Wuyi Rock Tea. It is a tea clone of a shrub. The tree's height is 1.6 meters, with an oversized crown and thick trunks. The branches and leaves are relatively crow. The leaf is smooth and oval-shaped. It is thick and crisp. The color is thick green. The leaf tip is blunt. The leaf margin has introversion angulation to the tile shape. It also has a high sprouting ability with plenty and small flowers.

　　成品茶特征：外形条索紧实、色泽青褐鲜润，香气浓郁高锐、有明显的桂皮香味，品质佳者带乳香。茶汤橙黄清澈，滋味醇厚甘爽，略带刺激性，叶底匀亮，呈淡绿底镶红边，冲泡六七次仍有"岩韵"的肉桂香。

　　The characteristic of Rou Gui Oolong Tea is as follows: tight and heavy tea leaf, blueish auburn and bloom color, sharp and robust aroma, and cinnamon's distinct scent. The higher quality tea also has a fragrance like frankincense. The tea soup is orange-yellow and clear. The taste is mellow and thick with slight astringency. Brewed leaves are bright and even. The green leaf still has a red margin. It still has a YEN flavor cinnamon fragrance after brewing six or seven times.

独具"岩韵"
Unique YEN Flavor

武夷岩茶的品质特点，众说纷纭。一般来说，不外乎是外形粗壮乌润，泡后呈现"绿叶镶红边"，香气浓郁，滋味甘醇等。上等的武夷岩茶贵在其所具有的天然真味，滋味醇厚，内涵丰富，即岩茶特有的韵味，也被称为"岩韵"。由于武夷岩茶生长于得天独厚的环境中，叶内所含物质丰富，而且武夷岩茶的采制十分考究，本身成茶就具有一种奇特的风格和美妙的韵味。要想鉴赏此茶的风韵，要像品茶的行家一样，准备一套特制小巧的茶具，泡上一壶，先嗅其香，再试其味，慢慢品赏，细细体味，花香如出自幽谷，雅趣盎然。

Scholars have had endless viewpoints about the quality characteristics of Wuyi Rock tea. In general, they are no more than sturdy and black-bloom appearance. The edge and body of the brewed tea leaves are bright red after brewing. Its fragrance is strong, and the taste is sweet and mellow. Higher-grade Wuyi Rock tea is valuable for its natural and perfect taste. The taste is mellow and thick with abundant inclusions. Its collecting and processing are elaborate. Wuyi Rock Oolong Tea has a unique style and extraordinary charm. We should prepare a small set of specially-designed tea sets and then brew some tea to appreciate this tea's unique charm. At first, smell the aroma, then taste tea soup. Taste slowly and enjoy the flower fragrance. We can feel its elegant taste during drinking.

图片由茶人魏荣南提供
Photos by courtesy of Wei Rongnan

武夷大红袍
Wuyi Robe Tea (Da Hong Pao)

贡茶披红袍
Tribute Tea Draped with Red Robe

武夷大红袍，产于福建省武夷山，是中国茗苑中的奇葩，是岩茶之王，更有"茶中状元"之称，堪称国宝。传说，天心寺的一位高僧用九龙窠岩壁上的茶树芽叶制成的茶汤治好了一位皇帝的疾病，这位皇帝便将自己身上所穿的红袍脱下，盖在茶树上以示感谢。此后，被红袍盖过的几株茶树被染，远远望去通树红艳似火，犹如披着红色的袍子，"大红袍"由此而得名。从此，大红袍便成了年年岁岁的贡茶。

Wuyi Robe Tea (Da Hong Pao), the rare flower in Chinese tea gardens the king of Wuyi Rock teas, has been cultivated in Wuyi city, Fujian Province. It also has the name Tea Champion. There is a legend that an ill governor of Congan County was miraculously cured with tea given by a monk from Heaven Heart Temple. The governor covered the tea tree with his red robe to show his gratitude; that is how the tea got its name. After that, tea trees were dyed red and in full bloom when looking from afar, like they were covered with red robes. Hence the name of the tea. Since then, Da Hong Pao Tea had been the tribute tea year after year.

采摘仪式
Plucking Ceremony

清明节前，惊蛰之日，大红袍树下将举行一年一度的采摘仪式。这个采摘仪式在武夷山由来已久，真正有文字记载是在唐代。这种习俗从唐、宋、元、明、清代代相传，在元代时曾达到鼎盛。

The annual plucking ceremony is held during the spring equinox season and before the Qingming Festival every year. This is a time-honored custom in the Wuyi Mountain area. The recorded history has more than one thousand years. This cultural custom became adopted in Tang, Song, Yuan, Ming, and Qing dynasties and kept a vital position in the Yuan Dynasty.

每年采茶时，在武夷山修建的一座御茶园内，由一位德高望重的茶农来主持采摘仪式，茶工、茶农、茶师和村民百姓都集聚到那里敲锣打鼓，抬着山神、水果、牲畜等贡品前来贡茶，并在口中齐喊："茶发芽！茶发芽！"久而久之便形成了这个武夷山特有的采摘仪式。

A venerable tea farmer presided over the Imperial Tea Garden's plucking ceremony in the Wuyi Mountain area. Tea workers, farmers, masters, and villagers get together to beat the drums there. Mountain deities, fruits, livestock, and other tributes were carried out as the tributes. They shouted, "sprouting" together. It would constitute such a tradition over time.

香气四溢
Suffusing an Exquisite Fragrance

武夷大红袍属于品质特优的"名枞"，石壁和岩间滴水的独特生长环境使其具有独特的药效和卓越的品质，更润生出其浓郁的桂花香气。成品茶香气浓郁，滋味醇厚，饮后齿颊留香，经久不退。

Wuyi Robe Tea (Da Hong Pao) is a famous rock Oolong Tea with superior quality. Water is dripping between the vertical wall of rock and stones. This unique growing environment can make tea with a particular function and excellent quality. The ever-increasing environment makes its Osmanthus fragrance stronger. The finished tea is with heavy aroma and a mellow and thick taste. Sweet aftertaste can last a long time.

白毫银针
White Tip Silver Needle

白茶珍品
Treasure Type among White Teas

白茶属轻微发酵茶，是中国的特产，已有上千年历史，原为北宋贡品。其最早出产在浙江湖州安吉大山坞茶场，一般地区并不多见。现主要产于福建省的福鼎、政和、松溪和建阳等县，台湾省也有少量生产。其主要品种有白毫银针、白牡丹、贡眉、寿眉等。白茶具有银白多毫，芽头肥壮，汤色黄亮，滋味鲜醇，叶底嫩匀等特点。尤其是白毫银针，全都由披满白色茸毛的芽尖制成，形状挺直如针，汤色浅黄，鲜醇爽口，饮后令人回味无穷，是白茶中的精品。

由于白毫银针中氨基酸的含量比普通茶叶高，而茶多酚又比普通茶叶低许多，且只能在每年春季采摘一次，因此产量稀少且价格昂贵。因白茶性温凉、健脾胃，具有退热降火等功效，一直深受我国港澳地区，东南亚和欧美等国家消费者的喜爱，所以早年间中国的白茶主要用于出口，在国内茶叶市场上较为罕见。

White Tea is a light fermentation and specialty tea of China with a thousand years of history in China which was initially a tribute tea in the Northern Song Dynasty. White Tea was originated in the Dashanwu tea plantation in Anji County, Huzhou City, in Zhejiang Province. It is rare in other areas. It is mainly cultivated in Fuding, Zhenghe, Songxi, and Jianyang County in Fujian Province. Few White Teas are produced in Taiwan Province. Its main varieties are White Tip Silver Needle, White Peony Tea, Tribute Eyebrow Tea, Long Brow Tea, etc. Its characteristics include plenty of silver fuzz, fat and tender, bright yellow color soup with mellow and thick taste, and tender and even brewed leaves. Mostly the White Tip Silver Needle Tea has covered the fuzz and tip. It is as straight as a needle. The tea soup is light yellow with a mellow and brisk taste with an endless aftertaste. It is a boutique White Tea.

图片由茶人林有希提供
Photos by courtesy of Lin Youxi

Its tea-polyphenol reduce by 50% with twice the amino acid content as common teas. Tea leaves only can be plucked once each year. Therefore, this tea is very rare at a high price. The character of White Tea is slightly cold and affects the spleen and stomach with digestion, reduced fever, and so on. It has been trendy among Hong Kong and Macao regions, Southeast Asia, Europe, the United States, etc. So Chinese White Tea is used mainly for exporting before. It is rare in the domestic tea market.

形如其名
Tea Appearance

白毫银针形如其名。正是因其芽头肥壮，芽长近寸，全身披满茸毛，色白如银，外形圆紧纤细如针，故而得此"白毫银针"的雅号。

The White Tip Silver Needle White Tea got its name from its appearance. This tea got its nickname from the below characteristics, such as bold and fat tip bud, one-inch length tip, covered fuzz body, silver-white color, round and tight, and slender like a needle.

白毫银针经冲泡后，稍许便可见针针直立，忽上忽下，竞相沉浮。茶汤呈浅杏黄色，清澈透亮，香气清鲜，闻来沁人心脾，品来毫香显露，醇厚回甘，其滋味因产地不同而略有差异。

When brewed with water, the tea leaves stands on the end of the cup, which float on the water. The tea soup is a light apricot-yellow color. It is clear with a fresh aroma and refreshing the soul. When you drink this tea, you can smell the fragrance of the tea fuzz. It is a mellow, thick, and sweet aftertaste. The taste varies with its different producing areas.

白牡丹
White Peony
White Tea

白牡丹茶的由来
The Origin of White Peony Tea

  白牡丹属白茶类，其形似花朵，绿叶夹白色毫芽，冲泡之后碧绿的叶子衬托着嫩嫩的芽叶，形状优美宛若蓓蕾初开，故名"白牡丹"。在民间还曾流传过这种茶树是由牡丹花变成的传说。

  White Peony Tea is one kinds of White Tea. It looks like a flower. The green leaves are clamped with some white fuzz tips. Those jade Green Tea leaves set off the tender buds. The shape is graceful and elegant. It looks like a blooming tea bud, hence the name. There is a beautiful legend that the tea tree was transformed from a peony flower.

图片由茶人林有希提供
Photos by courtesy of Lin Youxi

产地分布
Distribution of Producing Area

白牡丹茶属于白茶，为福建特产。1922年白牡丹茶创制于大湖镇地区，同年政和开始栽培，成为主产区。19世纪60年代，松溪县曾经一度盛产白牡丹茶，如今产区主要分布于福建的政和、建阳、松溪、福鼎等县。

White Peony Tea is one kinds of White Tea that is a unique ogricultural product the Fujian Province. White Peony White Tea was created in Fujian Dahu in 1922 and then cultivated in Zhenghe County in the same year. It has become the main producing area. Songxi County abounded with White Peony Tea in the 1860s. The production areas are mainly distributed in Zhenghe, Jianyang, Songxi, and Fuding County in Fujian Province.

图片由茶人杨丰提供
Photos by courtesy of Yang Feng

品质特征
QUALITY CHARACTERISTICS

白牡丹茶清淡高雅，冲泡后深绿色的芽叶抱着嫩芽，形似蓓蕾初开。上品白牡丹茶叶张肥嫩，毫心肥壮，叶缘微卷，叶背遍布白毫。茶汤清澈，橙黄或杏黄色，滋味甘醇清新，叶底浅灰，叶脉微红。常饮白牡丹茶，具有祛暑、明目、通血管、抗辐射及解毒等药用功效。

White Peony White Tea is plain, thin, and elegant. Its shape looks like a blooming flower after brewed with water. The leaves are fat and tender with a bold and silver heart. The leaves margin are sight curly, and its underneath is covered with white fuzz. The tea soup is clear, orange-yellow, and apricot yellow with a mellow and clear taste. Brewed leaves are light gray with a light red limb. It has the medical effects of reducing summer heat, benefiting the eyes, promoting blood circulation, anti-radiation and detoxifying the body, etc.

茉莉花茶
Jasmine Tea

人间第一香
A Gorgeous Fragrance in the World

茉莉花，原产自波斯，汉代传入中国，已有 1700 多年的历史。它是一种花色洁白、叶色翠绿，小花型的花卉，花小素淡，芬芳怡人。茉莉花兼有梅花的清芬、兰花的幽雅和玫瑰的甜郁，与兰花、桂花并称三大香祖，并享有"人间第一香"之美誉。

Jasmine is native to Persia and was introduced into China in the Han Dynasty of 1700. Its color is pure white with jade green. It is a kind of small flowers with hints of floral aromas. Jasmine has the advantages of plum blossoms, faint scent, orchids' elegant smell, and sweet roses. These three flowers varieties are the world's three strongest fragrance flowers. It also enjoys the reputation of the first fragrance in the world.

茉莉花茶，是在茶中加入茉莉花朵后熏制而成的，自古被视为窨花茶中之名品。在茶与茉莉花的窨制过程中，融茶的清香和花的芬芳为一炉，茶叶充分吸收花香的成分，使其既有茶香，又有花香，从而成为不可多得的茶中美味。茉莉花茶是花茶中的主要产品，历史悠久，备受欢迎，且被广泛流传。此茶的品质特点是茉莉花香浓郁、洁白宜人，冲泡和饮用过程中满室飘香，能够给人带来身心愉悦的感受。

Jasmine tea is made by adding jasmine flowers to tea, Its has been regarded as a famous product in flower tea since ancient times. The delicate fragrance of tea and the sweet-smelling of flowers mixed. Tea leaves can fully absorb the aromatic constituents to make Jasmine Tea with the flowers' aroma. It is a rare, delicious tea. Jasmine Tea is the primary type of Scented Tea with a long history. It is popular and widespread. The quality of this tea is characterized by the heavy flower scent of jasmine and its pure white color. During the brewing and drinking, the jasmine perfume the whole room and let the person's mind and body dulcify.

品饮的享受
Enjoyment of Drinking

茉莉花茶融合茶叶之味和鲜花之香于一体，品饮茉莉花茶，不如说是在品赏一件茶的艺术品。当茉莉花茶被拨入洁白如玉的白瓷盖碗中时，茶叶与茉莉干花飘然落下，看着片片香茶飞舞，可闻清香高远，可赏韵味雅致。茶叶的淡淡素香映衬着茉莉花的馨香，顿时让人感到神清气爽。

Jasmine Tea is mixed with the delicate fragrance of tea leaves and sweet-smelling flowers. Enjoy drinking Jasmine Tea like appreciating a tea artwork. When Jasmine Tea leaves are put into a delicate white jade porcelain cover bowl, leaves and jasmine flowers are falling. They are fluttering in the tea bowl. You can smell the pure and elegant fragrance. The plain aroma with the jasmine fragrance is refreshing.

品饮茉莉花茶不仅是感观享受，而且还是一种精神享受，正所谓"杯中清香浮情趣"。

Drinking Jasmine Tea is not only a sense of enjoyment but also pleasure of the mind. So-called fresh tea is the spice of life.

西南名茶
Famous Tea in Southwest China

茶造篇
TEA PROCESSING

西南茶区是中国的四大一级茶区之一，茶树品种资源丰富，有灌木型和小乔木型茶树。更难能可贵的是，在部分地区还生长着乔木型茶树，有些乔木型茶树的树龄甚至在千年以上。西南茶区的影响力不但在我国是巨大的，在世界上也是独一无二，这里是世界茶树的发源地。西南茶区出产的名茶有云南普洱茶、云南沱茶、滇红功夫茶、四川红茶、四川蒙顶茶，等等。早在一千年以前，这里的紧压茶就已经随着马队的铃声在茶马古道上流通到全国，甚至越过边境，到达周边和更远的国家和地区。西南名茶是中国茶叶走向世界的第一步。

The Southwest Tea Region is one of the four major tea regions in China. These tea areas are the first class tea areas in China. There are shrub and small tree species. Moreover, some arbor-form trees also grow up in this area. Some old arbor tea trees even had over a thousand years ago.

The effect of the southwest China tea cultivation area has been profound in China and is of great significance to the world. This area is one of the places where tea trees originated in the world. The famous teas which grow in the southwest China tea cultivation area are Yunnan Province Pu Er Tea, Yunnan Province Tuo Tea (bowl-shaped compressed mass of tea leaves), Yunnan Province Gong Fu Black Tea, and Sichuan Province Mending Tea, and so on. The Compressed Tea in this area had been sold out worldwide and in other countries and regions through the Tea Horse Road thousands of years ago. The famous Southwest tea is the earliest Chinese tea shipped worldwide.

西南茶区
Southwest China Tea Cultivation Area

茶造篇 TEA PROCESSING

区域范围
Area Coverage

西南茶区位于中国西南部，茶树原产地的中心位于神农架、武陵山、巫山、方斗山以西；大渡河以东；红水河、南盘江、盈江以北；米仓山、大巴山以南。茶区范围包括四川、贵州、云南中北部和西藏东南部等地。该地区以高原和盆地为主要地形，有较好的水热条件，是中国最古老的产茶区。

The Southwest China tea cultivation area is in Southwestern China. The tea plantation center are located in the Shennongjia Forest Area, Wuling Mountain Area, Wushan Mountain Area, west of Fangdou Mountain Area, east part of the Dadu River, the Hongshui River Area, Nanpan River, and north of Ying River Area, Micang Mountain Area, and south of Daba Mountain Area. Coverage includes Sichuan, Guizhou, Central and Northern parts of Yunnan Province, and southeast Tibet. This tea area is mainly landform with plateaus and basins, which has better water and temperature condition for cultivating tea trees. It is also the oldest Chinese processing tea area.

地理特征
Geographical Features

西南茶区地形比较复杂，主要集中在盆地和高原地区。海拔高低悬殊，气候差别较大，以亚热带季风气候区为主。其主要的气候特点是春季较为干旱，夏季闷热，秋季雨水较多，适合各种类型的茶树生长。

The Southwest China tea cultivation area has a varied topography. It has been concentrated in the basin and plateau area. There is a big difference between the altitude and temperature, and is cosist largdy of the subtropical monsoon. The main climatic characteristics are as follows. It is relatively dry during the spring and has a humid and wet climate in summer with more autumn rain. It is suitable for the growth of various types of tea trees.

茶树品种 Tea Varieties

西南茶区载培茶树的种类很多，有灌木型和小乔木型茶树，部分地区还有乔木型茶树。适宜的生态条件，使得出产的茶叶品种也很丰富，该地区主要出产红碎茶、绿茶、普洱茶和紧压茶等。

The Tea of wide varieties are cultivated in the Southwest China tea cultivation area. Some arbor-form trees are also growing up in this area. And the optimum ecological condition it also can enrich tea varieties with some main tea products, such as Broken Black Tea, Green Tea, Pu Er Tea, Compressed Tea, etc.

特产名茶 Specialty Famous Tea

古老的西南茶区地形各异，气象万千，有着丰富的茶叶种类。出产的名茶有云南的普洱、沱茶、滇茶、紧压茶、翠华茶、大白茶、苍山雪绿，贵州的遵义毛峰、都匀毛尖、雷山银球茶、云雾茶，还有四川的蒙顶甘露、蒙顶黄芽、竹叶青、峨眉毛峰、碧潭飘雪、茉莉清茶、三花茶、嘉竹茶、崃山茶等。

Different landforms presented a continually changing panorama. It has rich tea resources in the Southwest China tea cultivation area from ancient times. There are lots of famous teas such as Yunnan Province Pu Er Tea, Yunnan Province Tuo Tea (Bowl-Shaped Compressed Tea), Yunnan Province Gongfu

Black Tea and Compressed Tea, Cuihua Temple Tea, Big White Tea, Cangshan Mountain Snow Green Tea, Zunyi Fuzz Tip Tea in Guizhou City, Duyun City Tip Tea, Leishan Mountain Silver Ball Tea, Mist Tea, and Sichuan Province Mengding Tea Ganlu Tea, Mengding Yellow Bud, Light Bamboo Leaf Tea, Emei Mountain Maofeng, Bi Tan Piao Xue Tea, Jasmine Tea, Three Flowers Tea, Jiazhu Tea, and Xiashan Tea and so on.

六大茶山
Six Major Tea Mountains of Pu Er Tea

"六大茶山"指盛产普洱茶的六座古茶山，其中包括攸乐古茶山、革登古茶山、倚邦古茶山、莽枝古茶山、蛮砖古茶山和曼撒古茶山。最早记载于清檀萃的《滇海虞衡志》载，"普茶名重于天下，出普洱所属六茶山，一曰攸乐、二曰革登、三曰倚邦、四曰莽枝、五曰蛮砖、六曰曼撒……"六大茶山自宋朝开始闻名天下，是中国最古老的茶区之一。除攸乐古茶山外，其余的五大茶山都在今天的云南省勐腊县，因位于西双版纳澜沧江以北，史称"江北六大茶山"。一江之隔的江南也有六大茶山，即勐宋、南糯、勐海、巴达、南峤和景迈。澜沧江一带的气候和地理环境十分适宜大叶茶的生长，因此这里出产的普洱茶品质十分出众，自古以来便被作为贡茶专用，名气远播，时至今日仍享誉中外。

The six central tea mountains refer to the are six ancient tea mountains that abound with Pu Er Tea. Youle Ancient Mountain, Gedeng Ancient Mountain,

Yibang Ancient Mountain, Mangzhi Ancient Mountain, and Mansa Ancient Mountain. The earliest records were found in Dian Hai Yu Heng Record. The six central tea mountains were famous worldwide from the Song Dynasty and are one of the oldest tea areas in China. Besides the Youle Ancient Mountain, the rest of the five ancient mountains are located in Mengla County in Yunan Province. They were known as the Six Major Mountains in the north of the Yangtze River as their location. There are also six major tea areas just across the Yangtze River Meng Song tea area, the Nannuo tea area, the Menghai tea area, the Bada tea area, the Nanqiao tea area, and the Jingmai tea area. The unique climate and geographical environment of the Lancang River are suitable for the growth of Yunnan Daye Tea. So the Pu Er Tea from this area has outstanding quality and was regarded as a tribute tea from ancient times. Even to this day, it is enjoying an excellent reputation.

滇红功夫茶
Yunnan Province Gongfu Black Tea

大自然的恩赐
Nature's Precious Gift

滇红功夫茶,又称滇红条茶,属大叶种类型的功夫茶。该茶主要产于云南澜沧江沿岸的临沧、保山、思茅、西双版纳、德宏、红河等6个地区的20多个县域,是中国功夫红茶中的后起之秀。滇红功夫茶芽叶肥壮,金毫显露,色泽乌黑油润,滋味浓厚鲜爽,香气高醇持久等独树一帜的品质,都源于澜沧江水和两岸山峦的滋养。因此,当地人将滇红视为大自然对人类的恩赐。

Yunnan Province Gongfu Black Tea is also called Yunnan Province Leaf Tea. It belongs to a big-leaf species by Gongfu Tea. That tea is mainly cultivated over twenty counties in six cities of Yunnan along the Lancang River coast in Yunnan Province. They are Lincang, Baoshan, Simao, Xishuangbanna,

Dehong, and Honghe. It is a rising star tea species among the Chinese Gongfu Black Teas. Bud tea leaf is fat and bold with plenty of golden fuzz. Its color is dark and blooms with a rich and mellow taste. The aroma is high and mellow, which can last a very long time. The unique quality is formed with the most fertile nourishment from Lancang water and mountains. Therefore, locate people called this tea the extra gift from nature to humankind.

图片由茶人张成仁提供
Photos by courtesy of Zhang Chengren

图片由茶人张成仁提供
Photos by courtesy of Zhang Chengren

品质鉴别
QUALITY IDENTIFICATION

滇红功夫茶因采制时期不同，其品质也具有季节性的变化，一般春茶比夏茶略胜一筹，夏茶又略胜于秋茶。滇红香气浓郁，其香以滇西茶区的云县、凤庆和昌宁出产的为佳，滋味醇厚，回味清爽，香气高长且带有淡淡花香；而滇南茶区所产的功夫茶味道虽然浓厚，却略带刺激性。除季节与产地外，根据茶的条索、整碎、老嫩、净度、色泽等外形情况，也可以综合判断滇红品质的优劣。滇红功夫茶以一芽一叶为主制成，以条缩紧结、洁净齐整、金毫多显、色泽乌润者为好。

Due to the different plucking times, there are seasonal changes in Yunnan Province Gongfu Black Tea's quality. Usually, spring tea is better than summer tea. Summer tea is a little better than autumn tea. The aroma of Yunnan Province Gongfu Black Tea is heavy and strong. The best flavor of Gongfu Black Tea, Which tea is planted in Yun County and Fengqing County, and Changning County. They have a mellow and thick taste, a mellow and typical aftertaste, and a high and long fragrance with a mild scent. Tea is heavy and thick, produced in the south of the Yunnan Province Tea Cultivation Area, but it has astringency. We can appraise tea quality with leaf appearance features, friability, tenderness, neatness, color, and so on. Except for the season and producing area. Yunnan Province Gongfu Black Tea is processed with one leaf and one bud leaf. And the suitable material are tightly, bold, neat tea leaves. They are black-bloom colors with plenty of golden fuzz.

品饮的方式
Drinking Methods

滇红的品饮从使用的茶具来划分，可分为"杯饮法"和"壶饮法"两种。从茶汤中是否添加其他调味品来划分，又可分为"清饮法"和"调饮法"两种。中国北方绝大部分地区，品饮红茶都采用"清饮法"，不在茶中添加其他的调料。而在广东、福建、台湾等地，多以加糖、奶或柠檬切片调饮为主，使其营养更丰富，味道也富于变幻；在西藏、内蒙古等民族聚集地，"调饮法"则更为普遍，加入更加丰富的调配料可烹制出美味的酥油茶和奶茶。

Those drinking methods of Yunnan Province Gongfu Black Tea are often divided into various tea sets. It has two ways cup drinking method and pot drinking method. And it also can be divided into light drinking and flavoring drinking by adding other flavorings. The light drinking method is used for drinking Black Tea in most of northern China. Do not add any spices or flavors to tea soup. However, flavoring drinking is used primarily in the Guangdong, Fujian, and Taiwan provinces. They add sugar, milk, and sliced lemon to make it more nutritious and colorful taste. The flavoring drinking was daily in Tibet, Inner Mongolia, and other ethnic groups. They mixed rich ingredients with tea soup to cook delicious buttered milk tea.

蜚声国际
World-Renowned

滇红功夫茶,是世界茶叶市场上著名的红茶品种之一。滇红功夫茶是云南省传统的出口商品,主要出口俄罗斯、波兰、英国、美国等30多个国家和地区,深受国际市场的欢迎。

Yunnan Province Gongfu Black Tea is among the most famous Black Teas in the international market. It is a traditional Yunnan Province export commodity. It is mainly exported to Russia, Poland, Britain, the United States, and other Eastern European countries and Western Europe, North America, in more than thirty countries and regions. Therefore, it is well-welcomed in the International market.

云南普洱茶
Yunnan Province Pu Er Tea

独特的品种
Unique Tea Variety

普洱茶是在云南大叶茶基础上培育出的一个新兴茶种，原运销集散地在普洱市，故因此而得名，距今已有1700多年的历史。普洱茶的产区气候温暖，雨量充沛，湿度较大，土层深厚，有机质含量丰富。

Pu Er Tea is a new species of breeding tea based on the Yunnan Province large leaf tea species. Its distribution and distribution center was Pu Er City, So called Pu Er Tea. That was 1700 years ago. The producing environment of Pu Er Tea not only with a warm and humid climate from the abundant rainfall and deep soil with a high content of organic material.

普洱茶采用优良品质的云南大叶种茶树的鲜叶为原料，可分为春茶、夏茶、秋茶三个规格。春茶又分为"春尖""春中"和"春尾"三个等级；夏茶又称为"二水"；秋茶又称为"谷花"。其中以"春尖"和"谷花"的品质为最佳。普洱茶条形粗壮结实，芽壮叶厚，白毫密布，香气高锐持久，滋味浓郁并具有刺激性，茶汤橙黄浓醇，入口后略感苦涩，稍后便顿生高雅沁心之感，香气可比幽兰清菊，甘津持久不散，回味长久。普洱茶有散茶与型茶两种，型茶根据形状的不同可分为：沱茶、饼茶和砖茶等。

Those good quality fresh leaves of Yunnan Province Large Leaf tea trees are selected as the material for processing Pu Er Tea. It can be divided into three grades: spring, summer, and autumn. Spring tea also can be divided into spring tip grade, mid-spring grade, and ending spring grade. Summer tea is also called Ershui Tea, which means "second time plucking." Autumn tea is also called Guhua Tea. Spring Tip Tea and Guhua Tea are the best among these. Those outstanding quality Pu Er Tea leaves are bold and sturdy. Its bud is fat and thick with plenty of white fuzz. The aroma is high, sharp, and can last a long time. The taste is heavy and robust. The color of the tea soup is orange-yellow. It is mellow and brisk. The fragrance can be compared with orchids and chrysanthemums. You can feel elegant after a slightly bitter taste. The sweet aftertaste can last a long time. Pu Er Tea has two kinds of bulk tea and shaped tea. It can be divided into Tuo Tea (bowl-shaped compressed mass of tea leaves), tea cake, and brick tea according to their shapes.

普洱茶长期以来都深受国内外茶人的肯定，远销我国港澳地区及日本、马来西亚、新加坡、美国、法国等十几个国家。海外侨胞和港澳同胞更是将普洱茶当做养生佳品，对其格外青睐。

Pu Er Tea has been famous and recognized by customers at home and abroad for a long time. It is exported to the Hong Kong and Macao regions, Japan, Malaysia, Singapore, the United States, France, and more than ten countries. Overseas Chinese and compatriots in Hong Kong and Macan regard Pu Er tea as a good health product and especially favor it.

老班章生普
Lao Banzhang Unfermented Pu Er Tea

根据民间传说，老班章这片土地是属于老曼峨先民的，当阿卡人（今老班章人的祖先）来到这个地方时，老曼峨人已经种植了一些茶树，并把这片土地借给老班章的祖先居住，老班章人的祖先于公元1476年建立了老班章寨（ba jia pu 哈尼语）。

Lao Banzhang village, according to folk legend, belongs to the ancestors of Lao Man'E. When the aka people (the ancestors of the present Lao Banzhang village people) came to this place, the Lao Man'E people have planted some tea trees and lent this land to Lao Banzhang's ancestors. As a result, the ancestors of the Lao Banzhang people established the Lao Banzhang village (BA Jia Pu Hani Language) in 1476.

老班章的干茶芽头肥大，嫩茎粗壮，茶毫明显，色泽润亮，经过冲泡以后，伸展比例较大。用手抓干茶可明显感受到茶叶的重实，有压手感。原料优质且制作工艺精良的干茶，颜色鲜活，如宝光色、糙米色。热香有蜜、兰香，香气持久，冷香清爽清新。香气纯净，无杂异味，自然生态的蜜兰香悠扬持久，沁人心脾。口感浓醇，苦涩即化，回甘迅速，持久，体感强烈。茶汤入口瞬间，口腔就能受到苦味涩感的冲击，随着茶汤流动，口腔内和喉部开始有甘甜的感觉，随着茶汤被咽下，吸气时口腔和喉部会有明显的清凉感，随之而来的就是愈加强烈的回甘的感觉，多种感官被同时调动，给人一种鲜活的感觉。茶汤绿黄明亮，有明显折光。叶底肥壮，柔嫩。老班章茶与很多其他的山头茶最大的不同之处在于，品饮老班章的生茶在给口腔强烈的冲击后，让人们不喜欢的苦味和涩感能够非常快速地退去，而且消失得很彻底，回甘生津和其他的体感来得也更快、更持久。

The dry tea leaf of Lao Banzhang village has fat leaf buds, thick tender stems, apparent tea fuzz, and bright colors. With high-quality fresh leaves and superior production technology, dry tea leaves have fresh colors. The aroma is pure and clear. The natural fragrance is lasting and refreshing. The bitterness and astringency will melt immediately, and the sweetness will return quickly and have a strong sense. When you

图片由茶人李开华提供
Photos by courtesy of Li Kaihua

drink the tea soup, you will feel bitter and astringent. But there will be a sweet feeling in the oral cavity and throat soon. As the tea soup is swallowed, there will be an apparent fantastic feeling in the oral pit and throat, which followed by an increasingly strong feeling of returning to sweetness. A variety of senses are mobilized simultaneously, and giving people a fresh feel. Those brewed leaves are green and yellow, bright, with apparent refraction.

The aftertaste is the most significant difference between Lao Banzhang Unfermented Pu Er Tea and many other teas. Lao Banzhang Raw Pu Er Tea has strongly impacted the oral cavity. Some people don't like bitterness, but it can fade quickly and disappear completely.

普洱茶的冲泡
Pu Er Tea Brewing

传统的云南普洱茶是用云南大叶种晒青毛茶经过特殊工艺精制而成。味道醇厚，具有陈香，茶味不易被冲泡出来，所以必须用滚烫的沸水进行冲泡，但也不宜过沸，这样的水含氧量过少，影响了茶叶的活性。一般茶与水的比例掌握在 1:50，在水的选择上，选用纯净水、矿泉水和山泉水为佳。若是砖茶、饼茶则需拨开放置两周后再进行冲泡，茶的味道才会更好。

The traditional Yunnan Province Pu Er Tea is processed and refined from those large leaf sun fixation primary tea leaves. The taste is mellow, rich, and with old flavor. Its flavor cannot easily be brewed out. So we must use boiled water to brew it. The boiling water will affect water activity with low oxygen content, so it can't be used. The tea-to-water ratio of Pu Er Tea is between 1:50. It is better to use pure water, mineral water, and mountain spring water. The brisk tea and tea cake need to be spread out to lay up for two weeks. The taste will be better at that time.

在茶具选择上，由于普洱茶的浓度较高，故宜选用腹大的壶，这样可避免茶汤过浓，建议以瓷壶、陶壶和紫砂壶为首选。

We select the big-size teapot to brew Pu Er Tea. It can help to avoid making too strong tea soup. And we suggest brewing them with porcelain pots, pottery pots, and dark-red enameled pottery.

普洱茶的种类
Types of Pu Er Tea

普洱茶的种类划分有多种依据，但并不复杂。根据树种不同，普洱茶可分为乔木和灌木两种：乔木，即采乔木树叶做茶青，因叶片较大，又称"大叶茶"；灌木，即采用灌木树叶做茶青，叶片较小，也称"小叶茶"。

There are several classification for dividing Pu Er Tea into types. But it's not complicated. According to the different species of trees, it has arbor form trees and shrub trees species. Leaves from arbor form trees also can be processed as fresh material leaves. It is the so-called small-leaf tea.

根据制作和加工方法的不同，普洱茶可分为生茶和熟茶。生茶初制后需要经历大自然发酵，陈化数年后，茶性由刺激转为温和，方可饮用，简称"生普"；熟茶由人工发酵制成，茶性温和，制作完成后即可饮用，简称"熟普"。

根据存放方式不同，普洱茶有干仓和湿仓之分。干仓普洱是存放于通风干燥的仓库中，茶叶自然发酵的普洱茶，陈化10~20年的品质最佳。湿仓普洱通常放在湿度较大的地下室，空气中的水分可以加快普洱的发酵速度，陈化速度较干仓普洱更快。

Pu Er Tea can be divided into Pu Er Raw Tea and Pu Er Ripe Tea according to different processing methods. The character of Pu Er Raw Tea turns to mild from astringency. Therefore, it is known simply as Pu Er Raw

Tea. Pu Er Ripe Tea is processed under artificial fermentation. Its character is gentle. You can drink tea soup after completing processing. Pu Er Tea has dry and wet storage kinds according to different storage methods and conditions. Dry storage Pu Er Tea can be kept in a ventilated dry warehouse. The quality of naturally fermented Pu Er Tea is the best with ten to twenty years of aging. Wet storage kind Pu Er Tea can be stored in the basement with high humidity. Ambient moisture can step up its fermentation rate. Aging speed is faster than dry storage way Pu Er Tea.

根据普洱成茶的外形不同，还可以细分为扁平圆盘状的普洱茶饼，形似碗一般大小的普洱沱茶，砖头形状长方形普洱砖茶和未经压制的普洱散茶等。

The different shapes of Pu Er Tea can also be divided into flat disc shape Pu Er Tea Cake, Tuo Tea (bowl-shaped compressed mass of tea leaves), rectangle shape Brisk Tea, and Pu Er Bulk Tea without compressed.

普洱茶的功效
The Effects of Pu Er Tea

长期以来，普洱茶深受人们的喜爱，除了因其茶质优良，止渴提神，因为其具有特殊的药用功效。普洱茶的药理功用在古籍中早有记载，清人赵学敏《本草纲目拾遗》云："普洱茶性温味香，……味苦性刻，解油腻牛羊毒，虚人禁用。苦涩逐痰，刮肠通泄……"现代人尤其重视普洱茶减肥、降压、防癌及抗衰老等功效，饮用人群日趋广泛。

People have sincerely liked Pu Er Tea for a long time. In addition to helping produce saliva and slaking thirst with good quality, it also has unique medicinal benefits. Its therapeutic benefits were recorded in the ancient medicine book. The Compendium of Materia Medica Gleaning, written by Zhao Xuemin in the Qing Dynasty, said Pu Er Tea is mild with a bitter taste. It can help remove greasy but is not recommended for that poor fitness. It is bitter and astringent and dispelling phlegm. Modern people have focused on the extraordinary effects of reducing weight, lowing blood pressure, anti-cancer, and anti-aging and other mircaculous effects. Therefore, more and more people drink Pu Er Tea.

普洱茶的收藏
Pu Er Tea Collection

普洱茶与其他种类的茶叶不同，绿茶以新采新制的为好，红茶或乌龙茶最长存储1~2年的时间，而普洱茶有越陈越香的特点，可以存放很长时间，如果存储得当可陈化百年以上。普洱存放的时间越长久，味道反而越醇厚，价值也越高。除散茶外，紧压成型的普洱茶可以制成各种形状，小如丸药，大如巨型南瓜，方形、球形、饼形，匾额、屏风、一盘象棋、一幅浮雕均可由压制的普洱茶呈现出来。在现代工艺的包装下，普洱茶不但可以被收藏，还可以用于赏玩。有些人把收藏普洱茶比作收藏葡萄酒，称其为"可以喝的古董"。

Pu Er Tea is different from other types of tea. The better Green Tea is processed with fresh tea leaves. Black Tea and Oolong Tea storage time are one to two years. And Pu Er Tea has the characteristics of being more fragrant with more extended storage. It can be kept for a long time, even used to store for a hundred years. The taste is more mellow and heavier, with longer storage time. Therefore, its value will also be higher. Pu Er Compressed Tea can be processed into different shapes. They may be as small as a pill or as big as a giant pumpkin. They also have many shapes, such as square shape, spherical, pie shape, plaque shape, folding screen shape, chess shape, and cameo shapes besides the bulk tea. Pu Er Tea not only can be collected but also can be admired under the packaging of modern technology. Some people compare Collected Pu Er Tea to wine. They regard Pu Er Tea as an antique that can be drunk.

普洱茶的储存
Storage of Pu Er Tea

普洱茶和其他茶叶不同，可以存放很长时间。但并非所有的普洱茶都是存放的时间越久越好，其中一些也有一定的"寿命"。

Pu Er Tea is different from other teas. It can be kept for a long time. But not all Pu Er Tea is better for more extended storage. Some of them have a specific nature time.

由于氧气可以加速茶叶的发酵陈化，潮湿的环境容易使茶叶霉变，太阳直射会破坏茶叶中的营养成分，空气中的异味也极易被茶饼吸附。因此，普洱茶一般存放在通风、干燥、阴凉、无杂味的地方。数量较多时可以放在瓷罐或紫砂缸里保存。

Due to oxygen can accelerate the fermentation and aging process in the air, mildew could be broadly spread in a damp environment. It could also damage its nutritional content through sun exposure. In addition, various peculiar smells can be absorbed easily by tea cakes. Therefore, Pu Er Tea should be kept in a ventilated, dry, cold, and non-smelly place. Many tea leaves should be stored in a Porcelain jar or purple sand cylinder.

江北名茶
Famous Tea in the North of the Yangtze River

江北茶区是中国的四大一级茶区之一，也是中国最北的产茶区。在天气相对寒冷，气候变化相对明显的区域，成功地种植茶树，并且制造出品质优秀的茶叶，是一个成功的试验和创举，也是茶的发展史上的一个里程碑，更是茶文化普及的一个证明。江北名茶的种类也许相对少一些，包括传统的六安瓜片、信阳毛尖及名茶新锐的崂山茶等，但是这些名茶禀赋了独特的小气候环境，从而具备了其他名茶所不具备的优特点，使中国名茶的目录更为详尽，为中国茶文化的发展添加了浓墨重彩的一笔。

The tea area, located on the North bank of the Yangtze River, is one of the four most extensive first-class tea plantations. It is the most northerly tea cultivation area in China. Tea trees were cultivated in relatively calm regions and with a changing climate for processing the higher quality tea. It was a successful experiment and pioneering work. It represented a milestone in the long history of tea development and proof of tea culture's popularization. There are fewer famous teas in this area, such as Luan Leaf Tea, Xinyang Tip Tea, and Laoshan Tea. These favorite teas are cultivated in a unique environment and with special features. This tea enriches Chinese tea categories. It plays an essential part in the long history of Chinese tea culture.

江北茶区
The North of the Yangtze River Tea Area

区域范围
Area Coverage

江北茶区位于长江以北，秦岭、淮河以南，东自山东半岛，西达大巴山，包括甘肃南部、陕西西部、湖北北部、河南南部、安徽北部、江苏北部、山东东南部等地。地形复杂，气温较其他茶产区偏低，是中国最北部的茶叶产区。

The North China tea cultivation area is located on the North Bank of the Yangtze River and the Qingling Mountains–Huaihe River's southern region. This area is from the Shandong peninsula in the east to the Daba Mountain in the west. It includes the south of Gansu Province, western Shanxi Province, northern Hubei Province, southern Henan Province, north of Anhui Province, north of Jiangsu Province, and southeast Shandong Province in complex terrain with low temperatures. This tea cultivation area is also the northernmost tea processing area in China.

地理特征
Main Geographical Features

江北茶区的地形比较复杂，茶区土壤以黄棕壤和棕壤为主，土壤酸碱度略偏高，是中国南北土壤的过渡类型。茶区长年气温较低，四季分明，冬季时间较长，年平均气温为15℃～16℃，冬季最低温为–10℃，茶树很容易遭受冻害。茶区年降水量偏少，约800～1100毫米，而且分布不均，干旱时节需要借助灌溉，因此茶树新梢的生长时间比较短，采茶时间只有180天左右，产量较低。这是中国茶树生长条件较为不利的区域，但这并没有影响江北茶区的茶叶品质。

The Northern China tea cultivation area has a varied topography. Its main soil types are diverse. They are mainly latosolic yellow-brown soil and brown soil with a slightly high Soil PH Value. It has four distinct seasons. The yearly average temperature is about 15℃～16℃. The winter is long, with the lowest temperature of –10℃. Tea trees are easier to be damaged in freezing weather. The annual precipitation is slightly less, about 800～1100 millimeters. It is unevenly distributed. Artificial irrigation is also needed in dry periods. So the growth time for growing shoots is shorter. The plucking time is about 180 days with lower production. This area is not suitable for cultivating tea trees in China. But this is not affected the quality of tea in the Northern China tea cultivation area.

茶树品种
Tea Varieties

江北茶区茶树类型主要为灌木型中小叶群体，抗寒性较强。这类茶树树冠较矮小，自然生长状态下，树高通常只有1.5～3米，主干与分枝不明显，分枝密集，多出自近地面根茎处，茶树叶片小，其根系分布较浅，侧根发达。这也是中国栽培最多的茶树品种。

The tea tree's primary type is a shrub and small and medium-leaf species with relatively strong cold resistance. Its tree crown is less and short. Under the same natural growing state, the tree's height is about 1.5 to 3 meters. Branching is not apparent and tends to phalanx pattern. The leaf is small in size. Its root distribution is shallower, with a well-developed lateral root. It is also the most widely cultivated tea tree variety.

特产名茶
Specialty Famous Tea

江北茶区由于受地理环境的影响，茶叶产量相对较低，但仍然有优质的茶叶出产。陕西西乡的午子仙毫、汉水银梭、秦巴雾峰、河南的信阳毛尖、安徽的六安瓜片、舒城兰花茶、天柱剑毫、金寨翠眉、山东崂山茶等都是品质上佳的名茶。其中崂山茶因其翠绿的色泽和清爽的口感被称为"茶中新贵"，以盛产绿茶而闻名的山东省崂山区也因此有"中国江北名茶之乡"的美誉。

As a result of geographical conditions, tea yield is relatively low in the Northern China tea cultivation area. But there are some higher quality teas, such as Wuzi Mountain Tip, Hanshui Yinsuo Tip, Qinshan Mountain Mist in Xixiang County Shanxi Province, Xinyang Tip in Henan Province, Luan leaf in Anhui Province, Shucheng City Orchid Tea, Tianzhu Mountain Tip, Jinzhai Mountain Area Eyebrows Green Tea and Laoshan Green Tea. Laoshan Green Tea was called a tea star with its pure and fresh taste. In addition, the Laoshan area has a good reputation as the hometown of famous tea the North of the Yangtze River.

茶具篇
Tea Ware

水为茶之母,器为茶之父。

Water is the mother, and utensil is the father of a cup of tea.

茶经四之器

Chapter 4: Tea Ware

风炉（含灰承）：用铜或铁铸成，形同古代的鼎的样子，壁厚约三分，直径约九分，中间空约六分，用泥涂糊。炉有三只脚，脚上铸有古文字二十一个：一只脚上写有"坎上巽下离于中"；一只脚上写有"体均五行去百疾"；另一只脚上写有"圣唐灭胡明年铸"。三只炉脚之间有三个洞口，炉底下的一个洞用来通风漏灰烬。三个洞口写有六个字，分别是"伊公""羹陆"和"氏茶"，就是"伊公羹、陆氏茶"的意思。炉上有架锅用的垛，其内分为三格：一格上画有野鸡的图案，野鸡是火禽，此为离卦；一格上画有似虎非虎的彪，彪是风兽，此为巽卦；一格上画有鱼的图案，鱼是水虫，此为坎卦。"巽"主风，"离"主火，"坎"主水。风能使火烧旺，火能把水烧开，因此要有此三卦。炉身的装饰通常还有花卉、树木、流水及其他图案花纹等。风炉的炉身，有的用铁锻造而成，有的用泥土烧制而成。风炉的灰承，通常是一个有三只脚的铁盘，用以将炉身托起。

Brewing Ware The brazier is made of copper, often iron. Its shape is similar to a Ding. (Ding was a vessel to cook food and an important ritual object in ancient times.) Its wall thickness should be about three centimeter and nine centimeters in diameter. Its middle part is about six centimeters. The brazier is mainly cast in clay. Usually, three-foot embossed with twenty-one characters each. On one ley is written "Kan Shang Xun Xia Li Yu Zhong；On the other ley is written." Ti Jun Wu Xing Qu Bai Ji；On the other ley is written "Sheng Tang Mie Hu Ming Nian Zhu". There are three holes between those three legs. The hole on the bottom is used for ventilation and ashes removal. There are six Chiese characters around the hole. They are as follows Yi Gong, Geng Lu, and Shi Cha. We could explain the text's meaning as Yigong Geng and Lushi Cha, which means Yigong's thick soup and Luyu's tea. A ring foot at its bottom holds and supports the pots. Three compartments divide this leg. There are some magnificent pheasant, monster, and fish patterns on it. And there are three matching diagrams—some decorations on the brazier, such as flowers, trees, streams, and other patterns. The brazier body is forged with high-quality steel or clay. The iron pan with three-foot is to hold up the brazier.

茶器
Tea Utensil

"水为茶之母，器为茶之父"。对器皿的强调和要求正体现了茶人对品茗的完美追求。陆羽在《茶经》中便精心设计了烹茶和品茗的二十余种茶器。

Water is the mother of tea, and utensil is the father of tea. The emphasis on tea utensils has highlighted the perfect pursuit of tea. The Classic of Tea, edited by Lu Yu, elaborated on more than twenty kinds of tea utensils for brewing and drinking.

一般茶器需要兼具实用和美感的特性。从备水到理茶、置茶、品茗和清洁，每一个环节和步骤都要求配备有专门且精致的茶器。中国古代，茶器使用的精细过程，还蕴涵着丰富的文化思想和礼仪。

Those tea utensils are both aesthetic and practical. A delicate and specialized tea utensil is required for each processing step. In addition, it is full of cultural thoughts and etiquette.

精巧别致的茶器在招待宾朋的时候，既是一种感官享受，也表达了深厚的情感。从茶艺的角度来说，品茶是展演性的艺术享受，细致精巧的茶器在品茶过程中增添了许多雅致情调。

Those delicate utensils can create a colorful sensory experience by entertaining people who visit you and show deep and sincere affection. Drinking tea is a kind of artistic enjoyment from the tea ceremony's perspective. These exquisite tea utensils are better for a romantic ceremony.

茶具篇 Tea Ware

备水器

Water Ware

煮水壶

Boiling Kettle

煮水壶是用来煮开水用的泡茶辅助器具。陶制的煮水壶有保温的作用，属较佳材质。现代的煮水壶，通常会在壶底加一层保温材质以保持水温。在茶艺表演泡茶的时候，使用较多的有紫砂提梁壶、玻璃提梁壶和不锈钢壶等。

The boiling kettle is a kind of supplementary wares for brewing tea. The pottery kettle has an insulating effect with its better material, which has a layer of insulation material at the bottom of warming. Usually, we use the purple sandy clay teapot with a loop handle, a glass kettle, and a stainless steel pot during the tea art ceremony.

茗 炉
Tea Stove

  茗炉是用来煮烧泡茶水的炉子。为表演茶艺的需要，现代茶艺馆经常备有一种"茗炉"，炉身为陶器，下有一金属支架，中间放置酒精灯，点燃后将装好水的水壶放在"茗炉"上，可用来烧水或保持水温，用于表演使用。

  A tea stove is used for keeping water boiling. Most modern tea shops use a unique pottery tea stove with a metal stent for tea art ceremony. The furnace is burned with a spirit lamp. Then put a kettle on the stove for boiling water and keeping constant water temperature. This perfect teaware is convenient for you to perform well.

  另外，现代茶艺馆及家庭使用最多的是"随手泡"，也称电茶壶。它是用电来加热烧水，加热时间较短，水开后可以自动断电，方便快捷。

  An instant electrical kettle is used in the modern tea shop and family. It is an electric heating device. The time for boiling water is short. Use a kettle with an automatic cut-out to boil.

图片由茶人李廷怀提供
Photos by courtesy of Li Tinghuai

水 注
Water Vessel

水注即品茶时注汤用的汤瓶，又称为"茶瓶"或"汤提点"。一般是壶嘴细长、壶身较高的水壶。可盛放冷水，注入煮水器加热；或盛放开水，温具时用来注水或者等水温稍降后来冲泡茶叶。

A water vessel is a hot water pitcher and soup pot containing cold water. Its spout openings are mostly long and thin. It can be filled with cold water and heated by boiling water.

水注的完美运用增进了茶艺阳春白雪似的精巧韵味。

Yang Wanli is a famous poet of the South Song Dynasty. He also mentioned that the perfect use of small vessels could fully express the tea ceremony's charms.

水 方
Shui Fang (Water Vessel)

水方是用来贮存生水的泡茶辅助用具。陆羽的《茶经》中记载："水方以木、槐、楸、梓等合之，其里并外缝漆之，受一斗。"（注：一般水方与存贮淋注茶壶水的茶船都被称为"水方"，实际二者功用有别。）水方的出现增加了茶艺的精细与优雅。

A water vessel is an auxiliary tool for storing unboiled water.

The Classic of Tea from Lu Yu recorded that there are wood, elm, and Chinese catalpa Shui Fang. The paints are evenly on both sides of the water vessel. The emergence of a water vessel enriches the tea ceremony's artistic conception.

理茶器
Preparing Ware

茶夹
Tea Clamp

　　茶夹又称为茶筷，功用与茶匙相似。用于烫洗杯具和将茶渣自茶壶中夹出，有人也用它夹着茶杯洗杯，防烫又卫生。明代李贽写有隽永小品《茶夹铭》，其言"我老无朋，朝夕唯汝……夙兴夜寐，我与子终始"，素朴悠然的词句增加了茶夹的文化意蕴。

　　Tea clamp is also called tea chopsticks. Its function is similar to the teaspoon. It is used to warm cups and take out brewing leaves from the teapot. Some people use tea clamps to clamp and wash teacups. It has the advantages of a high anti-scalding effect and is sanitary. Li Zhi wrote in the Ming Dynasty that the Tea Clamp mentioned that I was lonely and could live with a tea clamp I was happy and sleep at night but with my son from the beginning to the end. These carefree and straightforward words have extremely rich cultural implications.

茶桨 Tea Paddle

茶桨是用来撇去浮于茶汤表面的茶沫的用具，尖端可用于疏通壶嘴。茶叶第一次冲泡时，表面会浮起一层泡沫，此时可用茶桨拨去浮沫。品茶时出现的泡沫不能用嘴吹或直接倒掉，而是也要用茶浆拨去。在娴静之间，更显高雅品致。

A tea paddle is used for skimming off the scum on the surface of tea soup. Its pointed end is used to dredge the spout. The tea soup has a layer of foam for the first brewing. You can use a tea paddle to skim off the foam instead of mouth-blowing or pouring it out gently for drinking. Between the demure, It looks so graceful.

茶针 Tea Needle

茶针即一根细长如针形的泡茶用具，故名为茶针，多以竹、木制成。茶针除去用来疏通茶壶的内网，保持水流畅，还可用于疏通壶嘴及茶盘的出水孔，以免茶渣阻塞，造成出水不畅。茶针精致与修长尖细的外形，加上疏通涤涤的用途，为品茶增进了清静舒畅、精巧雅致的美好体验。

Its shape looks like a long and thin needle and it is mainly made of bamboo and wood. People use it to dodge teapots and clean the spouts to pour a sufficient water flow. The elaborate tea needle has a beautiful shape and is convenient to use to increase visual enjoyment and immersion significantly.

茶刀通常在冲泡普洱茶时使用。取一下块饼茶放入茶荷后，用茶刀轻轻撬开，将敲下的碎片放入壶中，冲泡时更容易得到较浓的茶汤。由于茶叶种类不同，用茶刀时不必将茶敲打得过于细碎，以免产生较多粉末。用茶刀适度按压，舒活茶叶，利于茶香发散，茶韵浓烈。

A tea knife is used for brewing Pu Er Teacake in general. After taking one piece of tea cake into the tea holder and gently pry it with the tea knife. Then put those tea leaves into the teapot. It can make tea soup with a heavy flavor quickly. We don't need to cut the tea cake too small to avoid making more tea powders. Press tea leaves appropriately to awaken them. It is better for brewing out its fragrance and expressing its charms.

图片由茶人蔡万涯提供
Photos by courtesy of Cai Wanya

置茶器
Tea Setting Tools

茶瓮
Tea Urn

茶瓮是用于大量贮存茶叶的容器，通常为陶瓷所制。小口鼓腹，贮藏防潮。也可以用马口铁制成双层箱，下层放干燥剂（通常用生石灰），上层用于贮藏茶叶，双层间以带孔搁板隔开。经过茶瓮贮存的茶叶，可以保持茶叶口味长期不变，甚至增加茶叶的陈放韵味。

A container is usually made from pottery and porcelain and used for bulk tea storage. The big size pot body with a small mouth has the effect of being moisture-proof. We can also use a double deck to store tea traditionally made of tin-coated iron. Add a desiccant on the bottom and put leaves on the top. A perforated shelf is used to separate them. It can help to keep the flavor of tea and even can enhance its charm.

图片由茶人李廷怀提供
Photos by courtesy of Li Tinghuai

茶具篇 Tea Ware

茶 罐
Tea Caddy

作为备查器具的茶罐一般分为茶样罐和贮茶罐两种。茶样罐为泡茶时用于盛放茶样的容器，体积较小，约装干茶30～50克，贮茶罐（或叫贮茶瓶）为大量贮藏茶叶用，约能贮茶250～500克。为确保密封，应使用双层盖或防潮盖。贮茶罐一般为金属或瓷质，且造型美观多样、韵味雅致丰富。

As a reference equipment, tea caddy has two kinds of tea sample caddy and tea storage caddy. The tea sample caddy is a small-size and used for storage, holding thirty to fifty grams of dry tea leaves. A tea storage caddy or tea storage bottle can store about two hundred fifty to five hundred grams of dry leaves. We should use the double vacuum or moisture-proof lid to ensure a tight seal. Tea storage caddy is made of metal or porcelain generally. It also has a beautiful shape and elegant features.

图片由茶人李廷怀提供
Photos by courtesy of Li Tinghuai

茶匙 Tea Spoon

茶匙是一种长柄、圆头、浅口的小匙，用于将茶叶由茶样罐中取出，或在从茶壶内取出茶渣时使用，不可以沾水。茶匙多为竹质，如今亦有黄杨木质，一端弯曲。茶匙要求坚固有力，古代也有以黄金、银、铜制成。

The teaspoon has a round head and a long ladle, taking tea leaves from a tea storage caddy. Or take out brewed leaves from teapots. A teaspoon cannot be wet. Most of the teaspoons are made of bamboo. Boxwood is sometimes used today. One side is flat, and the other is convex. Some were made of gold, silver, and copper in ancient times.

茶则 Tea Scoop

茶则是分盛茶叶用的器具，一般为竹制。将宽一点的竹杆切开，利用竹管内部自然形成的节隔，可制作成茶则。此种宽的茶则是盛散茶入壶的用具。另一类茶则偏小，有的一端尽头稍微向上隆起，在茶道中用来将粉末茶盛入茶碗。

A tea scoop is used for dividing and holding tea leaves. They

are usually made of bamboo. Slightly cut open wide bamboo poles at the bamboo joint to make a tea scoop. We use the wide tea scoop for holding tea leaves to put them into the teapot. Another kind of tea scoop is smaller. They use another end to put tea powders and dust into the tea bowl.

茶漏呈圆形漏斗状，形制小巧，也叫做茶斗。一般泡茶所用的茶壶壶口皆较小，当用小茶壶泡茶时，可将其放置壶口，茶叶从中经过后缓缓漏进壶中，以防茶叶洒落到壶外。茶漏在茶艺表演过程中具有导引茶叶入壶的功用，具有优雅的动感和韵律。

The tea strainer is round and funnel-shaped. It is small and also called a tea hopper. This is because the part of the teapot opening is small in size. Put the tea strainer close to the teapot opening when brewing tea. Tea leaves will slip into the teapot slowly to avoid spilling. We can use a tea strainer to put tea leaves into the teapot. It has an elegant dynamic and rhythm.

茶 荷
Tea Holder

茶荷既可以观看鉴赏茶样的质色，同时也可用来置茶分样。需先将茶叶装入茶荷内，此时可将茶荷递给客人，鉴赏茶叶外观，再用茶匙将茶荷内的茶叶拨入壶中。茶荷的使用增加了品茗的观赏性和情趣。

Tea holders can be used to appreciate tea leaves and for setting and dividing. Firstly, those tea leaves must be put into the tea holder to show guests' appreciation. At this time use a teaspoon to put leaves into the pot. The use of a tea holder can give drinkers more aesthetic sentiment.

图片由茶人蔡万涯提供
Photos by courtesy of Cai Wanya

品茗器
Tea Savoring Utensils

茶 海
Cha Hai Pot (Fairness Mug)

茶海也叫做茶盅或公道杯，形状似无盖的敞口茶壶。茶海的容积要大于壶或盖碗，一般为瓷器、紫砂或玻璃容器等。从外观上分为无柄和有柄两种，有的还有内置过滤网。

Cha Hai is also called the Gong Dao Mug and Fairness Mug. Its shape looks like an uncovered pot without a handle. The volume of the Cha Hai is larger than the teapot or cover bowl made of porcelain, purple clay, and glasses. It is divided into two types, with the handle or not. Some pots even have an inner filter screen.

茶海的功用大致为盛放泡好茶汤，再分倒各杯，使各杯茶汤浓度相若，还可用来沉淀茶渣。当茶壶内的茶汤浸泡至适当浓度后，将茶汤倒至茶海，再分倒于各小茶杯内，这样可以均匀茶汤的浓度。于茶海上覆一滤网，可以滤去茶渣、茶末。

The function of this sharing pot can be roughly described as follows. It can hold tea soup for dividing. Tea soup in each cup will be almost the same as this method. Someone covers a filter net on the Cha Hai Pot to help remove tea residues and powders. Avoid creating unevenness in the gradation of tea flavor. All tea soup should be poured into the Cha Hai Pot and then separated into teacups.

闻香杯
Fragrance-smelling Cup

闻香杯顾名思义，即用来品闻茶香的专用杯子。它的容积与品茗专用的品茗杯相仿，但杯身细长而高，容易聚香。使用闻香杯时，将茶杯倒扣在闻香杯上，用手将闻香杯托起，慢慢地倒转，使闻香杯倒扣在茶杯上，稳稳地将闻香杯竖直向上提起（此时茶汁已被转移到了茶杯内），将闻香杯再次倒转，使杯口朝上，双手掌心向内夹住闻香杯，靠近鼻孔，闻茶汤留下的余香。

As the name suggests, the fragrance-smelling cup is used to smelling tea fragrance. Its volume is similar to the Sipping Cups. But it is easier to gather fragrance with its higher and slender body. First, invert the teacup onto a fragrance-smelling cup, then hold it by hand.

Turn over the fragrance-smelling cup steadily to switch it on the teacup. Then set the fragrance-smelling cup upright. The tea soup is removed from the teacup at this moment. Invert it again to move the fragrance-smelling cup upward for smelling. Breathe through the nostril to enjoy the fragrance of tea soup.

品茗杯
Tea-sipping Cups

一提到茶具，首先令人想到精致的茶杯。茶杯是品茗时的重要茶具。现在常用的品茗杯主要有两种：一种是白瓷杯；另一种是紫砂杯。茶杯以白底为佳，便于观察汤色。同时，高腹杯又比低平杯更容易品得茶香。

We have to mention the elaborate tea-sipping cup when talking about the teaware. The sipping cup is an essential ware for enjoying tea. Now, there are two main types of cups. One kind is a white porcelain cup. The other is the boccaro cup with ceramic whitewares inwall. We also use pure purple clay teacups. A cup with a white bottom is better for color observation. The cup with a high body is even better for smelling.

杯 托
Cup Saucer

杯托是用以承托衬垫茶杯的碟子。杯托的出现是饮茶习俗的普及和茶具装饰多样化的结果,杯托是整个茶具的配套器具。茶托一般与所托茶杯在质地上应保持一致,以体现协调之美。

A cup saucer is a small plate used for holding teacups. The cup saucer's appearance stems from the continuous popularization of tea-drinking customs and diverse decoration of tea sets. The saucer and sipping cup share the same material, reflecting the beauty of harmony with nature.

图片由茶人蔡万涯提供
Photos by courtesy of Cai Wanya

小茶壶
Little Teapot

顾名思义，小茶壶是与大茶壶比较而言相对较小的品茗器具。小茶壶在泡茶中的创制始于明代，一般做工精细，适合独啜或者作为功夫茶具组中的泡茶壶出现。

As the name suggests, this little teapot is relatively tiny. It originated in the Ming Dynasty and is known for its fine detail and craftsmanship. It is best for sipping and brewing Gongfu tea.

用小茶壶泡出的茶，味道格外甘醇芳香。明清时代以江苏宜兴的紫砂壶最为著名，如果是出自名家之手，甚至会四方争购，价比黄金。

The taste is extraordinarily sweet and mellow. The most famous ceramic teapot was made in Yixing County, Jiangsu Province, during the Ming and Qing Dynasties. There was a rush for the little teapots made by a master.

盖　碗
Cover Bowl

盖碗由盖、碗、托三部分组成，为现代茶艺最常使用的器具，清雅的风格能反映出茶的色彩美和纯洁美。在古代，盖碗的使用有讲究的礼仪，同时也是一种身份的象征。碗盖可以防尘、保温、闻香，还可以用来拂去茶沫。鲁迅先生在《喝茶》一文中曾这样写道："喝好茶，是要用盖碗的。"

The cover bowl is the most used utensil for the tea ceremony; consisting of three parts: cover, bowl, and saucer. Its elegant style can show its color and tea's natural beauty. In ancient times, there was an entire etiquette for drinking tea with a covered bowl. It was also a status symbol. The cover bowl has a dustproof feature and has dozen certain functions. It can help preserve heat, make it easy to smell and remove tea foam. Mr. Lu Xun wrote the book Drinking Tea. The covered bowel should be used for drinking high-quality tea.

杯托与茶杯、茶盖相互配合，也符合"天、地、人"三才的文化意蕴。茶盖上，谓之"天"，杯托在下，谓之"地"，茶杯居中，是为"人"。用茶托托住茶杯，既美观，又可避免端茶烫手或茶汤溢出沾染桌巾，饱含人文关怀。

The cover bowl is the most used utensil for the tea ceremony; consisting of three parts: cover, bowl, and saucer. Its elegant style reveals the vibrant colors and the natural beauty of the tea. In ancient times, there was an entire etiquette for drinking tea with a covered bowl. It was also a status symbol. The cover bowl has a dustproof feature and has dozen certain functions. It can help preserve heat, make it easy to smell and remove tea foam. Mr. Lu Xun wrote the book Drinking Tea. The covered bowl should be used for drinking good quality tea.

洁净器
Utensil Cleaning Tools

茶　船
Tea Boat

茶船形状有盘形、碗形，不但托放茶碗，茶壶也可放置其中，盛热水时供暖壶烫杯之用，也可用于养壶。当注入壶中的水溢满时，茶船可将水接住，避免弄湿桌面。茶船有竹木、陶、瓷及金属制品等。

Those tea boats are disc-shaped or bowl-shaped. They not only can be used to hold a tea bowl but also can hold a teapot inside. People use them for warming teapots and nourishing teapots when holding hot water. In addition, the tea boat's design can collect excess water to avoid getting the table wet. The tea boat is made from bamboo, pottery, porcelain, and metal.

茶 盘
Tea Tray

茶盘是用来盛放茶壶、茶杯、茶道组、茶宠乃至茶食等器具的浅底器皿。其形状根据配套茶具，可方可圆或作扇形。形式可以是抽屉式或嵌入式，既可以是单层也可以是夹层，夹层还可用以盛废水。

The tea tray is a shallow utensil designed for placing teapots, teacups and tea tools, tea figurines, pets, and tea snakes. Its shape is divided into matching tea sets. It can be square-shaped circular-shaped, or fan-shaped. There are two main types of tea tray drawer-type and embedded-type. It can be a single layer or an interlayer. The interlayer can collect the tea water.

茶盘的选材广泛，金、木、竹、陶皆可取。金属茶盘简便耐用，竹制茶盘清雅相宜，陶瓷茶盘精致讲究。放置茶壶、茶杯用的加彩搪瓷茶盘，也曾一度受到不少茶人的欢迎。有了茶盘的摆放使品茗活动能在一个更为洁净齐整的环境中进行。

The tea tray has a wide material selection range. Such as gold, wood, bamboo, and pottery. These materials provide the advantages of simplicity. ease of use, and convenient operation. have the advantages of simplicity, ease, and convenient operation. The bamboo tray is elegant, and the ceramic tray is exquisite. The polychrome enamel tea tray and teacup were once popular among people. Then tea can be enjoyed in a clean and neat environment.

水 盂
Tea Basin

与文房中的水盂稍有不同，文房中的水盂用于盛磨墨用水，而茶艺中作为茶具洁净器皿的水盂只要用来贮放茶渣和废水。水盂多用陶瓷制作而成，也有用玉、石、紫砂等材质。

The tea basin is slightly different from the water pot of paper stationery. The water pot is the container for the ink-stick water of paper stationery. The tea basin is used for collecting tea residues and used water. Tea basins can be made of ceramics, jade, stone, and purple sand.

水盂往往造型丰富、制作精细，纹饰细致精美，一度成为文人雅士赏玩的对象，认为其具有息心养神、滋益文思的妙处。

They exhibit many characteristics, including multiple categories in intense colors and shapes. It was a favorite item of those scholar-refined scholars at that time. They thought the tea basin had nourished the effect of helping people feel relieved.

茶 巾
Tea Towel

茶巾俗称茶布，主要功用为擦干茶壶，在品茶之前将茶壶或茶海底部残留的水擦干，也可用来擦拭滴落桌面的水滴。茶巾置于茶盘与泡茶者之间的案上，宜采用麻、棉等吸湿性较好的材质制作。同时，茶巾需手感柔软，花纹要柔和，同时也可以起到装饰的作用。

A tea towel, also known as a tea napkin, is primarily used for wipe the teapot. Wipe up the teapot's inside infuser and the tea pitcher's bottom part before drinking. It can also mop the spilled water and put it between the tea tray and tea maker on the desk. Using suitable hygroscopic materials, such as linen and cotton, is appropriate. The tea towel is natural colors and soft. It also plays a vital role in serving as decoration.

容　则
Tea Canister

容则是摆放茶则、茶匙、茶夹等器皿的容器，属于洁净器的一种。容则取"海纳百川，有容则大"的寓意，有包养天地的韵味。容则一般为筒状，用来安放茶则、茶夹等茶具，以木制、竹制居多，且造型古朴，纹饰精雅，彰显品茗的神韵，与茶匙、茶夹、茶针、茶漏、茶则一起被称为"茶道六君子"。

A tea canister is a container designed for tea tools, such as a tea scoop, teaspoon, tea clip, etc. They serve as a cleaning tool for tea utensil. The tea canister contains profound knowledge. The sea admits hundreds of rivers for its capacity to hold. It can nourish the whole world. A tea canister is a cylindrical structure to place the tea scoop, tea clip, and other tea tools. It is elegant with classic simplicity and decorations, mainly wood and bamboo. It could a lesisurely lifestyle and cultivate an elegant life attitude. It is practical with the teaspoon, tea clip, tea needle, tea strainer, and tea scoop. They are called Six Heroes of Tea Ceremony.

茶具的种类
Types of Tea Ware

茶具篇
Tea Ware

"器为茶之父",茶具的材质对茶汤的香气和味道有重要的影响,因此茶具多以材质的不同来进行分类。如今,最常使用和出现最多的茶具主要有紫砂茶具、瓷器茶具、漆器茶具、金属茶具、玻璃茶具和竹木茶具六大类。其中精致典雅、精品荟萃的紫砂茶具和晶莹细腻、端庄淡雅的瓷器茶具在众多种类的茶具中占最大的比重,得到饮茶人士的广泛青睐,被誉为茶具上品的"景瓷宜陶"。

The tea utensil is the father of tea. However, its material can affect the aroma-absorbing capability. So tea utensil is classified according to their different material quality. Today, the most commonly used are six main types: purple sand clay tea wares, chinaware tea wares, lacquerware tea wares, metal tea wares, glass tea wares, and bamboo tea wares. Of course, those tea drinkers favor exquisite purple sand clay and elegant porcelain tea wares. Jingdezhen porcelain and YiXing County pottery are renowned for producing the highest quality tea wares.

气韵独特的紫砂茶具
Unique Purple Sand Clay Tea Wares

与生俱来的特质
Outstanding Soil Specificity

经过科学分析，紫砂为多孔性材质，气孔微细、气密度高，这种特殊的结构使它具有良好的透气性和吐纳的特性。当紫砂器遇热时，气孔张开，将胎土内贮存之物吐出来，器具之内贮存是茶，便吐茶香；若贮存是油，就会吐油；久置不用，吸收了空气中的尘垢，就会吐尘垢。所以紫砂壶用来泡茶效果最好，且因为它的贮换功能，泡起茶来还会越用越好，久用后，以沸水注入空壶也会有茶香溢出。

除此之外，紫砂对于冷热骤变的适应性极强，即使是在寒冬腊月，置于温火上烧茶或注入沸水，紫砂器也不会因温度骤变而胀裂。

Based on scientific analysis, purple sand soil is a kind of cellular soil with high air tightness. Purple sand clay has excellent

纳福（吕俊杰）

air permeability and absorbability with its structure. Its stomata will open when it is heated. Then discharge its contents of the clay soil. Tea can give out an aroma. Some oil will leak out if you use it for storing oil. If you do not use it for a while, the purple sand clay pot will soon accumulate dirt and dust. So the purple sand clay pot is the best ware for brewing tea. And tea soup will be better with its storage and exchange function. Tea fragrance will emanate from the empty pot when poured into boiling water. Beyond that, purple sand clay is a highly adaptive soil. It is strong enough to resist and not burst with boiled water in a cold winter.

形
Shape

由于含砂量较低，紫砂的可塑性非常强，"方非一式，圆不一相"。历代紫砂艺人潜心研究各种物态，汲取了中国传统工艺品的艺术特点，创作出的紫砂器造型千姿百态，不拘一格。

With its low liquid limit, purple sand clay soil has strong plasticity. Its shape is changeable. The rich shapes collect artists' creative intelligence and breed Yixing County teapots' unique style and aesthetic content through the ages. They are varied and incredibly lifelike.

色 Color

紫砂的色泽属暖色系，内外均不上釉，显得古朴沉稳，清明淡雅。在1000℃~1250℃的窑火里，制作者匠心独运，可烧炼出数十种缤纷的色彩，大致可分成紫、褐、红、黑、黄、绿等色系，色泽多样，变化微妙。

Purple sand-clay soil has a warm tone. Neither inside or outside is glazed, unsophisticated, and straightforward. The teapot has a quiet charm of its own. It is fired in kilns with a temperature of 1000℃~ 1250℃. Those craftsmen show originality and craftsmanship by creating them with different colors. It can be roughly dirided into several colors: purple, brown, red, black, yellow, and green. They have various colors and subtle changes.

玉屏（吕尧臣）

紫砂器的颜色烧成之后不会褪色，经过泡茶滋润后更可呈现出温润光泽与自然平和的质感，与其他陶土混浊不清的色泽有着天壤之别。

The purple sand clay teapot will not fade over time. Instead, it can show a warm luster and a nature and peaceful texture after moistening the tea. As a result, they are different than other pottery clay

明清时代，紫砂茶具的制作达到前所未有的顶峰。越来越多的书画名家、文人墨客参与紫砂壶的设计制作，他们将文学、书法、绘画、篆刻等艺术融为一体，装饰于壶盖、壶柄和壶身，使紫砂茶具不但成为茶饮的载体，更成为富于内蕴的艺术品。

Production peaked at purple sand clay tea wares in the Ming and Qing Dynasties. More and more famous ancient writers, scholars, and calligraphers have participated in the design, and produetion. They showed a perfect union of craftsmanship with literature, calligraphy, painting, and seal cutting to decorate the pot lid, handle, and body. It is a carrier of tea drinking and art with a cultural connotation.

紫砂茶具的装饰方法有烧制之前的陶刻、堆绘、纹样装饰等手法，还有烧成后的装饰则有釉彩、抛光和金银丝镶嵌等装饰手法。

Its decorating technique is to combines carving, drawing, and patterning. After burning it will be decorated with colorful underglaze, glaze, and filigree marquetry buffing.

由于紫砂陶坯具有良好的可塑性，易于雕刻，无论草、隶、篆、魏碑、钟鼎铭文等各种书体，花鸟、山水、人物等国画白描，或配以文字，情趣皆备。闲暇之余，品茗赏壶，余甘盘旋于口舌之际，体会铭文刻画的意境，当无所他求。

Because of purple sand clay pottery has good plasticity. It can be engraved with all kinds of patterns more easily. Plegardless of cursive calligraphy, of ficial script, Wei Bei, Zhong Ding in scriptions and othter calligraphy, and white paintings of flowers and birds, landscapes, characters and other traoltvnal Chinese pantirgs they hare pictures with illustrated, Dr have both emotions and interests. Pots have lots of designs and information on them. Enjoy tea soup by appreciating those artistic conceptions of teapots.

茶具篇 Tea Ware

天际（吕尧臣）

艺术之美
The Beauty of Art

满足紫砂壶的艺术之美则要将形、神、气、态四者之美融为一体，集于一身。形，指作品的外部轮廓美；神，即作品所具备的神韵；气，即陶艺所蕴涵的和谐色泽美；态，即作品的高低、肥瘦、刚柔、方圆等各种姿态。一件上佳的紫砂作品能够抒发艺术的语言，表现出生动的气韵和强烈的艺术感染力。

A purple clay teapot shows a perfect union of shape, spirit, morale, and appearance. Shape beauty is the external appearance of beauty. Spirit is a romantic charm. Morale is the harmonious color and beauty contained in ceramicart. Apperance is the height, weight, rigidity, softness, and roundness of the work are rarious postures. An excellent purple clay teapot can express the art language to show artistic conception and strong appeal.

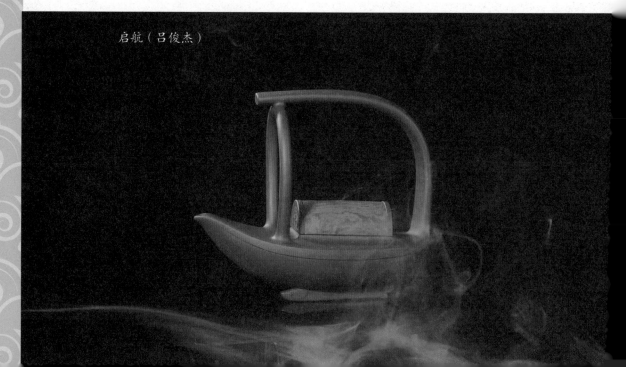

启航（吕俊杰）

紫砂壶的挑选
Selection of Purple Sandy Clay Tea Wares

作为一件完美的紫砂壶作品,既要具有实用性,又要兼备艺术价值。如果购买紫砂壶是为自用,那么选购者可以依据个人的饮茶习惯更多地考量其实用性;若是为了收藏,则要更多地侧重其工艺水平和艺术之美。

The teapot should have strong practicability and artistic value as a perfect purple sandy clay. If the teapot is just for personal use, it should adapt to its drinking habit. Instead, its craftsmanship and artistry can be emphasized and highlighted.

紫砂壶多是手工制品,一件壶要经过数十道工序才能制成。从百元到数万元,市场上紫砂壶的价值不一。界定紫砂壶的优劣,要看其是否手感舒适,整体协调。无论它是大是小,是曲是直,是否出于名家之手,只要能带来心灵的共鸣,愉悦身心,陶冶性情,就是一把值得珍爱的好壶。

Those purple clay tea wares are the most handmade work. A teapot should be made with dozens of processes. It has a different prices in market from one hundred to hundreds of thousands. Define the advantages and disadvantoges of the purple claypot, as long as it is comfortable to wse and overall coordinated. Whether it is large or small, curved or straight or from the hands of arenowned master, as long as it canbring lesorance to the soul, joy to the bocly and mind, and cuttivate character, it is a good pot with cherishing .

紫砂壶的收藏
Collection of Purple Sandy Clay Tea Wares

紫砂壶自明代出现以来就与茶文化紧密地结合在一起，除了其自身具有的实用价值和艺术价值之外，还有着丰富的文化内涵，是具有中华民族文化的外质内蕴的国粹。

Since the emergence of the Ming Dynasty, purple sandy clay teaware is tightly connected with the profound tea culture. Except its practical and with unique artistic charm and cultural value. It also has rich cultural conn otations. Purple sandy clay teaware is regarded as the quintessence of China.

温润细腻的瓷器茶具
Smooth and Delicate Porcelain Tea Wares

茶具篇 Tea Ware

中华瑰宝
Chinese Treasure

中国是陶瓷艺术的发源地,有着几千年的陶瓷发展史,早在宋代,就已呈现出瓷器制造业的极度繁荣景象。瓷器作为中国古代最著名的发明之一,是中华民族独有的珍贵文化遗产,为世界文明的发展作出了巨大贡献。

China is the world-famous birthplace of pottery with thousands of years of history. As early as the Song Dynasty, the porcelain manufacturing industry had already reached extreme prosperity. As one of the most well-known inventions, the chinaware is the Chinese nation's precious and unique cultural heritage. It also has significantly contributed to world civilization development.

唐代以来，陶瓷工艺被广泛地应用于茶具生产，作为历代茶具的上选材料，造出了许多传世的艺术精品。南越北邢、五大名窑、景德镇……一代代瓷器名家争奇斗艳，长盛不衰；青瓷、白瓷、黑瓷、彩瓷……各类制品色彩缤纷，形态各异。陶瓷制品独有的淡泊清雅提升了品茶情趣，深受好茶之人的喜爱。千百年来，经过不同时代、不同茶叶类型、不同饮茶习俗和方式、不同文化背景的考验，陶瓷始终贯穿于中国茶具文化的历史长河，至今仍保持着旺盛的生命力，成为现代普遍流行的茶具类型。

Chinese ceramic technology has been widely applied to teaware production since the Tang Dynasty. Created many legendary aitistic masterpieces, Many artworks were from the Tang-dynasty ceramic kiln in Hebei Province, Yue Kiln in Zhejiang Province, and the five famous kilns and Jingdezhen in Jiangxi Province. Generations of famous porcelain artists compete for beauty and their prosperity is mt good; porcelain, color porcelain, all kinds of products are color ful and different in shape. These unique and elegant porcelain tea wares can meet aesthetic interest. It is deeply loved by people who love tea. Through various tests, such as different eras, types of tea, tea-drinking customs and methods, and different cultural backgrounds, ceramic craftsmanship has run through the whole process of Chinese tea ware culture, ceramic craftsmanship has run through the whole process of Chinese teaware culture. China always maintains its development mode with thriving vitality. As a result, it has become the most popular type of tea county in modern times.

白 瓷
White Porcelain

白瓷于北朝时期已见雏形，但那时的釉色并不纯净，白中泛灰。到了唐朝，河北邢窑烧制出的白瓷则是自白釉瓷器出现以来的完美产物。它土质细润，坯质色泽纯洁，如霜胜雪。成品茶具轻巧精美，壁坚而薄，器型流畅，敲之音清韵长，传热、保温性能皆强。因色泽纯白光洁，能更鲜明地映衬出各种类型茶汤之颜色，适用范围较青瓷更广。杜甫曾有诗称赞白瓷茶碗："大邑烧瓷轻且坚，扣如哀玉锦城传。"

White ceramic production has seen its shape in the Northern Dynasties. Its glaze is not pure white exactly, but a bit gray at that time. By the Tang Dynasty, The white porcelain fired from the ceramic kiln in Hebei Province is the most perfect product since the existence of white glaze prolelain. Bullying Fkost and snow. clay is fine and glossy with pure color and high purity. Tea wares are light and exquisite with a solid wall. Its shape is smooth. You can even make the best music by hitting tea wares. White porcelain teaware has good diathermancy and heat preservation. Because the colon is pure white and smooth, it can dearly affect the color of various types of tea soup. Du Fu once wrote a poem praising the white porcelain tea bowl. Dayi porcelain is light and strong, like the mourning biography of Yujin City. It also has broader applicability than celadon. Du Fu once hada poem praising the white porcelain tea bowl: "Dayi porcelain is light and strong, like mourning Bioaraphy of Yujin Gty."

元代以后，江西景德镇的白瓷茶具较之前更上一层楼，它造型精美，装饰典雅，其外壁多绘有名人书画加以点缀，艺术气息更为浓厚，欣赏价值极高，堪称饮茶器皿之珍品并远销国外。至今白瓷依旧作为使用率较高的茶具，久盛不衰。

白瓷的出现除了打破青瓷的垄断地位外，另一深远意义则在于为后代茶具盛行的青花瓷、釉里红瓷、五彩瓷和粉彩瓷等彩瓷器打下了深厚的工艺基础，为陶瓷茶具的发展注入青瓷为玻璃质的透明淡绿色青釉，瓷色纯净，青翠欲滴，既清澈如冰，又温润如玉，制造出来的茶具质感轻薄，圆滑柔和，得到陆羽在《茶经》中的大力推崇。

The quality of white porcelain tea wares was better after the Yuan Dynasty. It was beautifully crafted and elegantly decorated. Their outer wall was decorated with famous paintings and calligraphy. They are very artistic with high appreciation value. It is a precious utensil for drinking tea and is exported to foreign countries. White porcelain tea wares are still the most commonly used utensils for drinking tea today. Long tasting prosperity. White porcelain broke the celadon monopoly and laid a solid foundation for the technique for the generations. Tea wares are light-textured, smooth, and soft. They have been widely advocated by The Classic of Tea, edited by Lu Yu.

黑 瓷
Black Porcelain

　　黑瓷茶具开始于晚唐，在宋朝初步发展后达到鼎盛，延续于元、明、清时衰微。宋代斗茶之风盛行，为黑瓷茶具的流行创造了非常便利的条件。白色茶沫与黑色茶盏色调对比分明，便于观察，且黑瓷胎体较厚，能够长时间保持茶温，最适宜于斗茶所用。因此，黑瓷茶具成为宋代瓷器茶具的主要品种。

　　Black porcelain tea wares began in the late Tang Dynasty. They reached their peak after the initial development of the Song Dynasty, continuing in the Yuan Dynasty and declining in the Qing Dynasty. Tea competitions prevailed in the Song Dynasty. It also created a suitable condition for the popularity of black porcelain tea wares. White color tea foam and black color tea wares are observed clearly. Those thick black porcelain tea wares can keep the temperature of tea soup for a long time. Therefore, it is better for a tea competition. Black porcelain wares had become the main tea ware variety in the Song Dynasty.

　　宋代有很多大量生产黑瓷茶具的瓷窑，其中以福建建窑生产的建盏最为著名。其自然烧制技法产生了显著的变异。符合宋代文人崇尚自然之美的审美价值观。它的深色与茶的绿色形成了完美的对比，也增加了一些审美情趣。建安兔毫盏代表了黑瓷茶具的最高峰。

Many porcelain kilns produced black porcelain tea wares in the Song Dynasty. The Jian Kiln in Fujian Province is the most famous. The natural firing techniques created a significant variation, in line with the Song Dynasty's scholarly values of appreciating nature's beauty. Its dark colors perfectly contrast the green or black of the tea they held. It also adds some aesthetic sentiment. Jianan County Hare's fur temmoku bowls represent the highest peak of ancient black porcelain tea wares.

仿故宫清代御用银兔毫茶碗（孙建兴）

到了明朝，因饮茶方式及茶叶类型的改变，黑瓷茶具不再流行，因此，它不得不从那时起退出了历史舞台。

Black porcelain wares were not popular in the Ming Dynasty due to their drinking style and tea variety. Therefore, it had to withdraw from that stage of history from then.

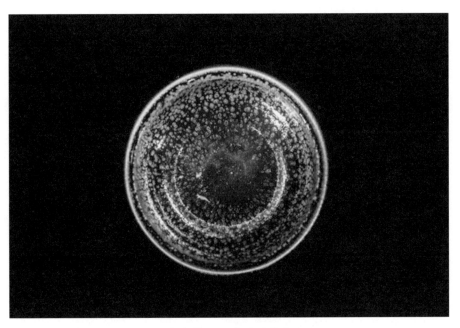

油滴盏　江雪（孙莉）
图片由茶人孙建兴及孙莉提供
Photos by courtesy of Sun Jianxing and Sun Li

彩　瓷
Decorative Porcelain

　　彩瓷是指带彩绘装饰的瓷器，比单色釉瓷更具美感，更有艺术性。可细分为釉下彩、釉上彩、釉中彩及釉上、釉下相结合的斗彩，于明清年间兴起。景德镇最负盛名的青花瓷茶具是典型的釉下彩瓷代表，其胎质坚硬，釉面光润明亮，釉色晶莹，蓝白花纹相映，淡雅清幽，造型多样。清代出现的釉上彩充分吸收了中国绘画的表现方式，瓷面上的绘画图案更富层次，出现凹凸浓淡的变化，立体感强，光泽透亮，粉润柔和。

Decorative porcelain is a kind of chinaware with decorative painting. It is more beautiful and more artistic than the monochrome glaze. The decorative porcelain is subdivided into the under-glazed, on-glazed, in-glazed decoration, and glazed decoration combined with both. It began to rise during the period of the Ming and Qing Dynasties. The blue and white porcelain teaware in Jing de zhen is typical under-glazed porcelain. Its tire quality is hard with a smooth and bright glaze. The glaze color is glittering, and the blue and white stripes complement each other, making it elegant and serene with diverse designs. The craftsmanship appearing in the Qing Dynasty of on glazed color fully integrated the performance of the painting. The painting pattern has abundant gradation to produce a powerful sense of stereo. It is glossy and transparent.

图片由茶人李廷怀提供
Photos by courtesy of Li Tinghuai

骨 瓷
Bone Porcelain

骨瓷属软质瓷，是以骨粉加上石英混合而成，是当今世界上公认的高档的瓷种。其制作过程极为复杂，工艺特殊，标准严格。然而，骨瓷起源于英国，现在中国也已经掌握了骨瓷的制作工艺并能生产出在工艺制作、花面设计、器型风格上独具中华特色的骨瓷茶具。

Bone porcelain is a kind of soft porcelain made of bone meal and quartz, the world's most recognized and most delicate porcelain type. Its creation process involved a unique technique and strict standards. However, bone porcelain originated in the UK. China has mastered bone porcelain and patterning design equipment style with solid Chinese cultural characteristics and national style.

骨瓷茶具将使用和艺术的双重价值集于一身，以自身独有的雍容典雅，成为收藏名家视之如珍的首选，无愧为当世"瓷中之王"。

Bone porcelain has both good application value and artistic value. It has become the first choice for viewing it as precious for collectors and deluxe banquets with its uniqueness and elegance. It is worthy of the name king of porcelain.

景瓷器
Jingdezhen Porcelain

景德镇陶瓷作为茶具上品，应用最广。景瓷兴盛于宋，北宋时期，景瓷即以青出于蓝、灵巧秀丽的青白釉茶具天下闻名；到了元明清时期，景德瓷艺发展迅速，广征博采，大胆创新，被称之为"至精至美之瓷，莫不出于景德镇"。它在国际上因"薄如纸，白如玉，明亮如镜、声音如钟"而被世界所熟知，景德镇瓷器历来被全球的古董瓷器收藏家们所重视。

As the highest grade material of teaware, Jingdezhen porcelain is the most widely used. It sprang up in the Song Dynasty and the Northern Song Dynasty and spread worldwide for its green and blue color and delicacy. During the Yuan, Ming, and Qing dynasties, It had developed rapidly with bold innovations. All elaborate and delicate porcelain came from Jingdezhen at that time.

景瓷瓷质优良，造型精美，色彩丰富，装饰多样。在装饰方面有青花、釉里红、古彩、粉彩、斗彩、新彩、釉下五彩、青花玲珑等，其中尤以青花、粉彩产品为大宗，有色釉为名产，标志着景瓷打破以前单一色调，开辟了彩瓷新时代。釉色品种很多，有青、蓝、红、黄、黑等类，每一大色系釉彩还分出更细致的小类。将彩绘人物、山水、花鸟都用于釉面装饰，为茶具增添了人文艺术气息，承载了中国传统文化精华。所有这些创造，在中国瓷器历史上都"开创了一代未有之奇",成为中国陶瓷艺术的杰出代表。

It has excellent quality. exquisite design. rich colovs, diverse decorations and opened up a new era of color porcelain. There are many types of color introductions. induding blue, red, yellow, and black each maior color series is also divided into more detailed srlb cateqaies. Among them, blue and white, pink products are the main products, and colored porcelain is the famous product. making the break from the previous single tone and opening up a riew era of colored povcelain. Porcelain is decorated with colored drawings such as characters, scenery, and flower-and-bird to be artistic. They added to the scenic, culturally solid presence. All these creations opened up a new era of China porcelain and became an outstanding representative of ceramic.

茶具篇 Tea Ware

淳朴典雅的漆器茶具

Elegant and Straightforward Lacquer Teawares

主要产地
Main Producing Areas

漆器的历史悠久，据史料记载，早在夏禹时代已见广泛使用。在汉代，漆器被作为日用器具已逐渐普遍，后经历朝蓬勃发展，漆器制成茶具，有据可考的始见于清代。现代漆器工艺广泛分布于山西平遥、甘肃天水、陕西凤翔及北京、江苏、上海、重庆、福建等地。其中，以风格富丽华贵的北京雕漆、以镶嵌螺钿为特色的扬州漆器和色泽光亮、轻巧美观的福建脱胎漆器最富特色，制成的漆器茶具受到人们的广泛喜爱。

Lacquerware has a long and colorful history. According to the historical record, it was widely used in the Xia Dynasty. Lacquerware was more frequently used for home appliances in the Han Dynasty.

It flourished from dynasty to dynasty, was created into tea wares, and then started to show in the Qing Dynasty. The craftsmanship is widely distributed in Pingyao Shanxi Province, Tianshui Gansu Province, Fengxiang Shanxi Province, Beijing, Jiangsu, Shanghai, Chongqing, Fujian, etc. Those elegant Luxury Beijing Carved Lacquer, Silver-Inlaid Yangzhou Lacquerware, and Fujian Bodiless Lacquerware with bright color, light weight and beautiful appearance are the most characteristic and precious tewares. They are very popular with people.

独具特色
Unique Style

漆器茶具是用漆涂在各种茶具的表面而制成的饮茶器具。漆料可以涂在各种材质的表面，常见的有陶器、瓷器、铜器、竹木器具等。漆器完好地保留了原有材质保温性能好、造型多样、使用灵巧方便等作为茶具的优秀特质。

Lacquer tea wares are to coat those kinds of wares with lacquer. The lacquer is painted on the surface of different material wares. Pottery, porcelain, copper, and bamboo wood wares are the most common. Lacquer wares can maintain excellent quality with suitable heat-insulating properties, diverse modeling, and light and convenient use.

绚丽的工艺
Magnificent Craft

自从漆器出现，历经三国的兴起，两汉的鼎盛，唐宋的发展和明清的全盛，漆器的制作工艺不断地丰富和进步。艺人们将化学工艺与手工技艺结合起来，创造出许多制作和装饰漆器的工艺技法，种类繁复，水平高超，影响深远。

Lacquerware tea sets had grown in the Three Kingdoms Period (220–265), developed in the Tang and Song Dynasties, and flourished in the Ming and Qing Dynasties since they emerged. Its manufacturing craft had been greatly improved by the rapid development of science and technology. Those artists combined chemical technology with handcraft skills to create countless decoration crafts and techniques. The variety is diveise, the level is superb, and the impart is profound.

华贵不实的金玉茶具

Gimcrack Precious Stone and Metals Tea Wares

茶具篇 Tea Ware

尊贵的金银器具
Valued Gold and Silver Utensils

在唐代,金银已经被用于茶具的制造。陕西法门寺出土的宫廷御用金银茶具,工艺精巧绝伦,形美质佳,极尽皇家奢华,代表了当时金属茶具工艺的最高水平。如果说唐代金银茶具还只是多为上流社会权贵富人显示身份地位所用,那么到了宋代,金银制造工艺则又有进步,加上民风奢靡,斗茶之风盛行,金银茶具更被视为上品,即使在民间茶肆也有使用。蔡襄的《茶录》中曾有记载,当时流行的斗茶用具均以黄金为上,次一些则以银铁或瓷石为之,充分说明了金银器具在当时的受重视程度。金银茶具绚烂夺目,在工艺上穷极精巧,自身造价极高,无论何时都是奢侈的高档消费品。

Gold and silver were already beginning to create tea wares in the Tang Dynasty. Many of the unearthed royal palace gold silver tea wares were excavated from the underground palace of Famen Temple in the Shanxi Province. They are so famous for their exquisite craftsmanship and excellent quality. These extravagant wares represented and reflected the highest level of craftwork of that time. The upper class only used the gold-silver tea wares to show status or success in the Tang Dynasty. Then the manufacturing process of precious stones and metals made more progress in the Song Dynasty. Tea competition was extravagant and flourished during that period. Those gold-silver utensils were regarded as the top-grade wares used in the teahouse. The most popular wares are made of gold in the book of Tea Record from Cai Xiang. The slightly worse tea wares are made of silver iron and porcelain stone. The importance of gold and silver utensils has been fully proved by both the past and the present. Valued gold and silver tea wares are brilliant with exquisite craftsmanship and high cost. So it is the luxury consumer good in every dynasty.

图片由茶人林陵祥提供
Photos by courtesy of Lin Lingxiang

独具优势的锡器
Unique Tea Tin Wares

　　金、银、铜、锡都是制造茶具的可选之材。中国锡制茶具的兴起始于明末清初，明代许次纾的《茶疏》中曾说："锡备柔刚，味不咸涩，作铫最良。"锡铫煮水，可令水快速沸腾，无熟汤之气，饮茶极佳，口感别具清香。锡铫被推崇，应是锡之提炼技术精益的结果。以前曾有人认为用锡制茶具煮水泡茶易夺香败味，使茶味走样，其实主要是因锡质不纯所致。

　　Generally, these materials can be selected as optional for producing tea wares such as gold, silver, bronze, and tin. Tinware originated in China during the late Ming and early Qing Dynasties. The Ming dynasty scholar Xu Cishu wrote in his book Tea Shu. Tin is neither hard nor soft, the best material for boiling water utensils. It boils water much faster. Its fresh and pure taste is good for tasting teas. Those tin utensils are an advance of refining technology. It is thought that tin wares would easily cause off-flavor and weak tea taste. The main reason is that tin tea wares have high impurities.

　　锡材较软，延展性好，易于加工，可以做成各种茶具。但锡器的最大优越性在于贮茶。锡制茶罐具有良好的密闭性和透气性，储茶可使茶叶色味不变，长久保持茶叶自身的鲜翠芬芳，自古就被推为贮茶之极品，为其他材质所不及。

锡器外表美观鲜亮，雅致肃静，又不失高贵的大家之风，具有很高的艺术欣赏价值，既是实用器具，也是收藏上选。

Tin is soft with good flexibility, which is easy to process. It can be made of various teawares. The most significant advantage of tin wares is good storage. The tin canister has good closeness and air permeability. Storage of tea with utensils can keep tea leaves color and taste. It can maintain the fragrance of tea leaves. Tin was promoted as one of the best materials for tea storage. Modern tin teaware has fine artistry and a beautiful, well-designed. It has both high artistic and crucial economic value for collection.

玉石茶具
Jade Tea Wares

自从陶瓷茶具出现，玉石茶具就被逐渐淘汰并取代。究其原因，除了陶瓷本身具有适宜泡茶的特点之外，陶瓷工艺的迅速发展也是一个方面。当然最重要的还是玉石本身，因材料稀少且价格昂贵，加上雕琢困难，用玉石制成的茶具成为普通人可望不可及的奢侈品。也正因如此，自唐宋以来，使用玉石茶具饮茶，仅仅成了贵族富人炫耀财富与地位的方式之一。

Since ceramic tea wares emerged, jade tea wares were gradually phased out and replaced. The main reason is our rapid development of ceramic technology with attractive features. By its nature, those jade tea wares can be used for brewing tea several times. Of course, the most important is the jade material. Its high price is with its rare reserves.

Moreover, carving, cutting, and polishing a jade is challenging. Jade tea wares have become a pipe dream for ordinary people. As a result, jade tea wares became one way to show off their wealth and status as early as in the Tang and Song Dynasties.

然而玉石茶具并没有在茶文化的历史中销声匿迹，只是扮演着极为次要的角色而始终存在着。时至今日，仍有一些现代人对玉石茶具格外青睐。材料上乘、工艺精湛的玉石茶具，外观精美，雕琢精细，平滑圆润，光泽流转，是价值不菲的工艺品，值得赏玩与收藏。

However, jade tea wares did not disappear in the history of tea culture. On the contrary, it only acts as a minor role. Even now, jade tea wares are still popular among some modern people. Good material and sound craft with an elegant appearance in lustrous colors are valuable artifacts for collection and appreciation.

由盛及衰
From the Flourish to Decline

金玉茶具在不同的历史时期都曾有过流行，古人因不同背景的茶艺文化，对金玉茶具的褒贬不一。明朝张谦德就曾在所著《茶经》上，把金、银茶具列为次等，而铜、锡茶具则又属下等。金玉茶具对于品茶而言并无显著优势，金银富贵之光与茶艺文化朴素自然、精行简德的人文主义思想也难以契合，没有对茶艺起到太大的推进作用，且由于材质、做工各方面原因，其造价昂贵，不利于广泛普及。

The precious gold and jade tea wares have been popular in every period. People pass different judgments on gold and jade teawares according to their different background cultures. Zhang Qiande listed gold and silver tea wares as second-rate. Those bronze and tin were the lower quality in his version of The Classic of Tea. There are no significant advantages to drinking tea with gold and jade wares. But those Luxury tea wares are irreconcilable with the natural and straightforward tea culture. It is not conducive to popularization for various reasons, such as its material and artistry.

明代之后，随着饮茶方式及品茶风气的改变创新，以及瓷器和紫砂器的勃然兴起，金玉茶具的实用价值越来越弱，渐渐为人们所弃用。即使在现代，金玉茶具的收藏价值依旧大于实用价值，更多的是作为一种富贵的象征，远不及陶瓷及紫砂茶具生命力之旺盛，在中国茶具文化历史上注定会昙花一现，终究未成主流。

Teawares are an essential part of tea history. And it was changing with the drink styles in every historical period. The practical use of gold jade tea wares is lower with the change and innovation of the tasting methods. Its collection value is more significant than practical use, even in modern society. Those gold and jade tea wares have been regarded as a symbol of wealth. The vitality of purple clay sand tea wares is much stronger. It was doomed to be a historical flash and has never become mainstream.

通透夺目的玻璃茶具
Translucent and Attractive Glass Tea Wares

古之琉璃
Ancient Colored Glaze

陕西法门寺地宫出土的皇室茶具中，已有琉璃茶盏和茶托，虽然造型还很简一，质地也略显浑浊，透明度较低，但却是茶具文化考古的一个重大发现，证明了琉璃茶具最晚始于唐代，是现代玻璃茶具的始祖。

Some colored glazed tea bowls and cup saucers were unearthed from the Famen Temple in the Shanxi Province. Though only structure and turbid material, it was a significant discovery in tea wares culture and archeology. It proved that the glazed tea wares were introduced no later than the Tang Dynasty.

不可取代的优势
Unique Advantages

近现代，随着玻璃制造工艺的发展，古之珍贵的琉璃终于发展成为今天价廉物美的玻璃，并以自己的特点和优势成为茶具选材中的后起之秀，开始走进寻常百姓家。玻璃质地完全透明，光可鉴人，传热快，不透气。其可塑性极大，制成的茶具形态各异，外观秀美，晶莹剔透，光彩夺目。

用玻璃茶杯泡茶时，茶汤色泽由浅入深的晕染，澄清碧绿；茶叶由团至展的沉浮飘动，如繁花盛放；杯中热气升腾缭绕，点滴细微的变化皆可通过透明的玻璃杯被茶人尽收眼底，将冲泡茶叶的整个过程转化为无尽的视觉动态之美，观之赏心悦目，别有一番情趣。尤其是冲泡各种名茶时，柔嫩细软的茶叶之美在玻璃杯中得到淋漓尽致的发挥，不负名优茶品珍贵的品赏价值。

With the development of the glazing technique in modern times, those unique glaze has developed into inexpensive and high-quality glass. It has become a rising star with its features and characteristics to get into ordinary citizens' lives. It is entirely transparent and brilliant. There are some effects of fast heat transferring speed and right tightness. It can be produced in different shapes with extreme plasticity. They are elegant, clear, and bright.

The tea soup is jade green from light to dark when brewing tea with a glass cup. Tea leaves are floating and unfolding. They are just like blooming flowers. Steaming a hot cup of tea can make you energetic. It created a unique and exciting experience. The scenic beauty of tea leaves gets incisively and vividly performance. It has an appreciating value.

中国古代的琉璃制作技术起步很早。唐代时，中西方文化交流十分频繁，在西方琉璃器不断传入的影响下，国人也开始烧制琉璃茶具，即玻璃茶具。

The making technology of glaze has begun exceptionally early. There was frequent contact with the western world during the Tang Dynasty. While the Western glaze was continuously introduced, we created our glazed tea wares.

由于原料容易获得，制作成本低，工艺简单，因此玻璃制品的价格十分低廉。又兼玻璃杯泡茶，便于观汤色、赏茶舞，给人们带来品茶之余的视觉享受，因而备受人们的喜爱。从古至今，从南至北，从富贵人家到寻常百姓，玻璃茶具日益得到广泛的使用。

Glass is easily obtainable with low cost and simple techniques, which are very cheap. It is better for observation to give the perfect visual enjoyment. So it can make us feel very welcome. Glass tea wares are increasingly being used in countries from the earliest to the present day and south to north.

自然粗犷的竹木茶具
Natural and rugged Bamboo Tea Wares

茶具篇
Tea Ware

留传于民间
Inherited from the past

竹木茶具起源于民间，作为中华传统文化的珍宝，其经济实惠的特点最易被大众所接受。唐代茶圣陆羽在《茶经》中所列的整套28种茶具，多数也为竹木所制。历史上，茶文化遍及乡野，在广大农村，很多劳动人民都使用廉价的竹碗或木碗喝茶。在南方海岛之地，还有人用椰壳加工而成的茶具泡茶，取材更为随意，清新别致，更像一件艺术欣赏品。

Bamboo teaware originated in folk customs as a treasure of traditional Chinese culture. The economic and practical wares were the best for our ordinary working people. Lu Yu edited The Classic of Tea, which recorded that most of the twenty-eight kinds of tea tools were made of bamboo. Tea culture has spread all over the country in history. The great majority of the rural people drank tea from cheap bamboo bowls or wooden bowls. Some people even use coconut husk to make tea wares in the south. Those unique wares are a work of art.

茶具的选用
Tea Wares Selection

茶具篇
Tea Ware

因时制宜
Adjust to the Changing Social Situation

茶具随着中国茶文化史的发展而演变改进。饮茶方法不断进步，茶具也随之向更为精良、实用的方向一点点发展变化。在元代之前，喝茶以煮饮的方式为主，茶具主要有煮水的鼎镬和饮茶的杯盏；唐宋时期陶瓷工艺发达，青白瓷茶盏鼎盛一时；宋代偏好斗茶，黑盏占据主导，至元明时期又被淘汰；元代之后，煮茶法逐渐被泡茶法所代替，煮水器具主要改为汤瓶，然后用壶来泡茶，茶壶越来越受到人们的重视；紫砂壶在明清两代广受青睐，至今仍多为沿用；清代，盖碗异军突起，在北京及四川一带的茶馆里开始盛行。

The evolvement and improvement of tea wares are intimately associated with the history of Chinese tea culture. With the constant progress of drinking tea's rapid development, tea ware has become more

347

excellent and practical. Brewing methodology is the primary drinking method before the Yuan Dynasty in history. In ancient times, a cauldron was used as a cooking vessel and drinking cup. The ceramic technique was advanced in the Tang and Song Dynasties. Bluish white porcelain tea wares flourished at that time. People preferred tea competition in the Song Dynasty. Black glazed teacup was the most popular but had become obsolete in the Yuan and Ming Dynasties. After the Yuan Dynasty. The hot water pitcher gradually replaced the heating water ware. The brewing method was gradually reylaced by the brewing methad People more and more valued the teapot. The ceramic teapot had broad appeal in the Ming and Qing Dynasties, still used today. The tea bowl with a fitted cover emerged unexpectedly in the Qing Dynasty. It became popular in the teahouse around Beijing and Sichuan Province.

到了近现代，除了紫砂壶、陶瓷茶具、盖碗这些从古代流传下来的茶具之外，玻璃杯作为新兴的茶具，更有助于观察名优细茶的形色，也在茶具市场中占据了一席之地。

In addition to those purple sandy clay pots and the ceramic tea wares and cover bowls from ancient times, the glass tea wares help observe the shape and color of famous tea leaves as a new type of tea set in modern times. It also occupies a place in the tea wares market.

因人而异
Vary with Each Individual

茶具是品茶人在日常生活中不可或缺的用品，自然要根据品茶人自身不同的审美观点、品饮习惯、职业、年龄、性别、性格等特点，挑选出最合自己心意，使用起来最得心应手的茶具。

Tea wares are indispensable utensils in everyday life for drinkers. Tea drinkers will select their handy tea wares according to their aesthetic viewpoint, drinking habit, position, age, gender, and personality.

对于喜爱古典文化的饮者，古色古香的紫砂器具自然最受青睐。紫砂壶造型古朴，色泽典雅，看上去稳重而不张扬，泡出的茶叶茶香味浓，贵在味道，适合中老年人选用。仿古造型的瓷器也能为茶平添几分古典韵味。而时尚的年轻人对茶更多的要求是香清味纯，重在外观，宜用高身瓷壶、瓷杯，或直筒玻璃杯沏茶，看上去简练快捷，更富现代气息。若品茶人是文人雅士，则切不可用大壶大杯，而要用小巧玲珑的紫砂或瓷质壶杯。提壶在手，慢冲慢酌，才能营造出幽雅清闲的氛围。

The antique ceramic tea wares are the most popular for those who love classical culture. The ceramic teaware has an ancient shape and elegant color. They are skillful, generous, and steady. Tea aroma and taste are strong and loved by aged people. Those imitative antique porcelain tea wares can add gracefulness, delicateness, and gentleness. It is also good teaware for aesthetic appreciation for young fashion guys. It looks simple and is enriched with a strong sense of era and modernization. A few refined scholars prefer to use small and exquisite ceramic tea wares. It can create a quiet and elegant environment.

与茶相宜
Best Match

"壶添品茗情趣，茶增壶艺价值。"好茶与好茶具，恰如红花绿叶，配合好了必能相映生辉。

A good teapot can make the brewing step more amusing, and then good tea leaves can add value to the teapot. Red flowers look more beautiful against green leaves. This is because they have enriched each other.

不同的茶，有不同的冲泡方法，不同的观赏特色，不同的汤色、香气和滋味，不可一概而论，不分种类就随手拿一套茶具应付了事，那样绝对体会不到冲泡和品饮的乐趣。对于绿茶、黄茶、白茶、花茶，应采用自然冲泡法，以透明玻璃杯为佳，可将茶汤色泽及茶叶舒展起伏的姿态一览无余。对于红茶、黑茶来说，一般采取功夫茶冲泡法，以闻香品味为首要，而观形略次，可以把紫砂壶和瓷杯结合使用。紫砂泡茶，有助于掌握茶叶的冲泡时间，紫砂多含气隙，吸收茶汤精华，有助于提升茶的芬芳；瓷杯饮用，有助于衬托深色茶汤，更好地观察茶汤本色，也便于事后清洗。浅斟慢饮，与神而会，虽可提神醒脑，其实早已沉醉，身边一切俱已空灵。

There are different brewing methods for different tea types. You cannot randomly select any tea wares for brewing. Transparent glass is suitable for brewing Green Tea, Yellow Tea, White Tea, and Scented Tea. You can appreciate the elegant gesture of brewing leaves and tea soup color when brewing it. Black Tea and Dark Tea usually are brewed with the traditional Gong Fu Tea brewing method. Smelling is the essential step, and then appreciating the appearance comes second. The ceramic teapot and porcelain cup need to be used together. Making tea with the purple sandy clay teapot can help you know the brewing time. There are many air gaps of purple sandy clayware for absorbing tea soup's essence and enhancing its fragrance. It can also help set off the dark-colored tea soup with a porcelain cup and is easy to clean. Sipping and enjoying tea soup slowly can help refresh yourself and show its nature in the spiritual space.

茶艺篇

Tea Arts

茶艺是茶的综合艺术，茶艺是一种文化。茶艺，萌芽于唐，发扬于宋，改革于明，极盛于清，可谓有相当的历史渊源，自成一脉。茶是中国人日常生活中不可缺少的一部分，饮茶习惯在中国人身上根深蒂固，已有上千年的历史。

Tea art is a comprehensive art of tea, and tea art is a form of culture. Tea art originated in the Tang Dynasty, developed in the Song Dynasty, reformed in the Ming Dynasty, and flourished in the Qing Dynasty. It has a considerable historical origin and has its system. Tea is an indispensable part of Chinese daily life, and the habit of drinking tea is deeply ingrained in Chinese people, with a history of over a thousand years.

茶经五之煮

Chapter 5: Tea Brewing

煮茶的水，以山泉水为最好，其次是江水、河水，井水最差。山泉水最好取用乳泉、池塘等流动缓慢的水，不要饮用瀑布、涌泉之类奔流湍急的水，长期饮用这种水会使人的颈部生病。由数条溪流汇合，蓄于山谷中的水，虽然清澈澄净，但因一直不流动，从酷暑到霜降期间也许会产生污秽的物质和毒素，取用时要先挖一处缺口，使腐水流出，同时新的泉水涓涓流入，这时的水才能汲取饮用。取用江河的水，要到距离人群远的地方去取。井水则要在很多人汲水的井中汲取。

Springwater is the best water type for brewing tea. Second, we select the river water. And the well water is the worst for tea. The slow-moving water in the stalactites spring and pond are the best types. Do not use the rushing falls water and slow-moving fountain water. People would have some neck diseases with this water for an extended period because the valley water is not flowing. It has some dirty and toxins things. So this part of the water needs to be drained away. At the same time, the fresh spring water could flow out. They can be a drink. The river water needs to be collected far away from the crowd. Then the excellent well water should be collected from the frequently used well.

水 Water

"水为茶之母",正是无水则无茶之意。水与茶依依关联,茶色、茶香、茶味都通过水来体现,可以说"清水出茶心",所以择水便理所当然地成为饮茶艺术中的一个重要组成部分。

Water is the mother of tea. Those characteristics of tea get a full embodiment, which can be described as the pure tea heart. Good tea should be brewed with good water. So, water selection has naturally become an essential part of the art of tea drinking.

自古,中国人就十分讲究饮茶择水,水中溶入了茶的清香芬芳,也融入了茶道的精神品质、文化意蕴和审美理念。烹茶鉴水,也是中国茶道的一大特色。

The Chinese are very particular about drinking tea and selecting brewing water from ancient times. Tea soup not only has a fresh fragrance.

The tea ceremony's cultural implication and aesthetic concept can all be experienced. There are many methods to distinguish water from ancient times to the present. Thus it can be seen that good water plays a vital role in brewing quality tea.

讲究的泡茶水
Fastidious Selection

天　水
Heaven Water

在茶道中，大自然的雨水、雪水、露水等被称为天水。天水是大自然给予人类的恩赐，自古以来是泡茶的上乘之水。古时工业不发达，环境没受到污染，天空明澈纯净，雨水、雪水也要比今天所见的洁净得多。因此食用雨水、雪水是常见的现象，既有益健康，又体现了茶道中人与自然的和谐统一。

唐代大诗人白居易《晚起》一诗写到"融雪煎香茗"，曹雪芹也在《红楼梦》中也不吝笔墨地描述了用隔年雨水和梅花上的雪水泡茶之韵。

In tea lore, the rain, snow, and dew in nature are called heaven water. Since ancient times, Heaven water has been a precious gift to human beings from nature and the best water type for brewing tea. In industrially backward China, the environmental quality was good. Rain and snow water was cleaner than now. So people can drink rainwater and snow at that time. Their tea ceremonies were good for health and fully reflected the harmony and unity between man and nature. Bai Juyi, the famous Tang poet, once wrote pottery. They melt snow to brew tea. Cao Xueqin also used some appropriate words to describe it. Some drinkers would use last year's rainwater and snow water to brew tea. That was recorded in his book A Dream of Red Mansions.

地 水
Earth-Water

地水指的是大自然中的山泉水、溪水、江河湖海之水和地下水等。陆羽曾在《茶经》中明确指出："其水，用山水上，江水中，井水下。其山水，拣乳泉、石池漫流者上……其江水，取去人远者。"他认为用山泉水泡茶最佳，因为山间的溪泉含有丰富的有益于人体健康的矿物质，为水中上品。江、河、湖水均为地表水，长年流动，所含矿物质不多，因受污染较重，反而含有较多杂质，混浊度大，用以沏茶难以取得较好的效果。但在远离人烟，植被生长繁茂之地，污染物较少，这里的江河湖水经过澄清后也不失为泡茶的好水。

The earth's water includes mountain spring water, stream water, rivers, lakes and sea, and underground water. The Classic of Tea recorded that the mountain spring water was the best for brewing tea, which Lu Yu edited. The mountain stream water was top grade with higher mineral content.

Rivers, lakes, and seawater could not reach the perfect brewing effect with its continually flowing and less mineral content.

软 水
Soft Water

泡茶用水有软水和硬水之分，所谓软水是指每公升水中钙离子和镁离子的含量不到10毫克。近代科学分析证明，一般在无污染的情况下，自然界中只有雪水、雨水和露水，即前文介绍的天水，才称得上是纯软水。虽然雨水在降落过程中会碰上尘埃和二氧化碳等物质，但含盐量和硬度都很小，历来被文人和茶人所喜爱，常用之煮茶。

The brewing water can be both soft and hard. The soft water contains no more than ten milligrams of Calcium ions and Magnesium ions per liter. Only snow, rain, and dew water exist in the non-pollution natural world. As mentioned earlier, heaven water is only pure. Even though rainwater would meet some dust and carbon dioxide while falling, its salinity and hardness are low. Therefore, it has been regarded as the best type of water for brewing.

硬 水
Hard Water

每公升水中钙离子和镁离子的含量超过 10 毫克的即为硬水了。除去天水外，其他如泉水、江水、河水、湖水和井水，这些地水都属于硬水。水的硬度影响水的 pH 值，而 pH 值又影响茶汤色泽，pH 值高会使茶汤颜色变深，茶中的茶黄素流失。水的硬度还会影响茶叶有效成分的溶解度。

A liter of hard water contains more than ten milligrams of calcium and magnesium ions. Springwater, river water, lake water, and well water are all hard water, except heaven. The hardness of water can affect the pH value. Then the pH value can affect the color of the tea soup. It will become darker with a higher pH value, and the Theaflavins also would lose. The hardness of water also affects the solubility of active ingredients. Hard water contains more Calcium ions, magnesium ions, and minerals.

泉 水
Spring Water

泉水水源多出自深山，或潜埋地层深处，而流出地面的泉水，经多次渗透过滤，水质一般比较稳定，以至有"泉从石出清宜冽"之说。在茶圣陆羽眼中，山泉水是地水中最适合泡茶的美水。天然泉水经沙土的过滤而具有较高的矿物质，大多数都有益于人体健康，并且因为流动而溶有丰富的气体，具有很强的活性。现代科学认为，泉水的水质清澈，因其从地下涌出时会经地层反复过滤，并在二氧化碳的作用下，溶入岩石和土壤中的矿物元素，同时，泉水在流动的过程中又吸收了新鲜空气，所以，这样的水具备了矿泉水的营养成分。然而，有的泉水由于溶解矿物质较多，致使含碱量和硬度不符合标准。所以，也并非所有的山泉水都能泡茶，有些泉水甚至含有硫黄等非饮用成分。

The most famous spring water comes from those remote mountains or deep ground. The river springs are clear and cold, with multiple filtration times. The saint of tea Lu Yu named the mountain spring water the most suitable for brewing tea globally. The natural springs have a higher level of beneficial minerals through the filtration of sandy soil. Most of them are good for health and have vigorous activities with flowing properties. Modern science assumes that clear spring water is filtered and purified with layers of earth several times. The action of carbon dioxide to dissolve in mineral elements from rock and soil. At the same time, the spring water absorbs fresh air during its flow. So this kind of water has the nutrition ingredients of mineral water. But some spring water cannot meet the standard with its alkali content due to too much-dissolved minerals. So, not all spring water can be brewed with tea. Some spring water even has non-drinkable components, such as sulfur.

天下名泉
Famous Springs in China

中冷泉
Zhong Leng Spring

后唐名士刘伯刍喜品天下甘泉，将适宜煮茶的泉水分为七等，中冷泉从众多名泉中脱颖而出，被品评为天下第一。中冷泉自此名扬天下，"天下第一泉"的美誉流传至今。

The celebrity Liu Bochu loved to taste all spring waters in the Later Tang Dynasty (923–926). He divided those springs which were suitable for brewing tea into seven grades. Zhong Leng Spring stood out and was rated the No.1 in China. Then it became famous since then. Its good fame came down the ages.

茶艺篇 Tea Arts

玉 泉
Yu Quan Spring

虽然陆羽和刘伯刍都喜爱品评天下名泉,且各自有心目中的第一泉,但还有另一个"天下第一泉"是因为得了乾隆皇帝御口亲封,名号也更为响亮,那就是位于北京西郊玉泉山东麓的玉泉。

Lu Yu and Liu Bochu would like to taste all the world's famous spring waters and have their ranking. The most famous spring is located in the western suburbs of Beijing. Emperor Qianlong gave him the title personally. That is Yu Quan Spring.

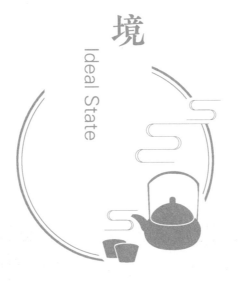

境
Ideal State

茶艺篇 Tea Arts

品茗的乐趣在于茶，也在于境。古往今来的爱茶之士无不在茶境上下足了功夫。茶境既包括品茗时秀美素朴的风景和清雅别致的气氛，也意指恬淡疏阔、宁静悠然的心灵之境。或禅韵，或清流，或楼台，或田园的物境；或超脱，或诗意，或幽远，或闲适的情境。物境与情境二者天然结合，便构成了几乎完美的品茶境界。曼妙的音乐，纵情的书画和寂静的棋局对弈，在如画的茶境之中，又增添了几分雅致。从本质上说，茶境是生活基础上的一种形式和追求。自然与心境在个人的品茗之中合而为一，直达物我两忘。完美的茶境，能够给人以独特而高雅的精神享受。

The joy of drinking tea is not only about tea but also the excellent atmosphere. Tea drinkers need time and energy to create an ideal state for enjoying tea. The perfect condition includes beautiful sceneries, an elegant atmosphere, and a leisurely pace. Zen rhyme, clear stream, terrace, and

other fields to be sublimated to the poetic realm. They also create the perfect state for enjoying tea. The charming music and wonderful calligraphy and silent chess game can impart a certain elegance to the environment. Essentially, drinking tea is a pursuit of pleasure, facilitating the emergence of life and tea. It constructs a pure ambit. The ideal state can give people unique and elegant spiritual enjoyment.

品茗佳境
Excellent Environment for Drinking Tea

茶艺篇 Tea Arts

佛门清净地
Pure Lands in Buddhism

　　茶与禅素来有缘，所谓"茶雁添诗句，天清莹道心"。早在魏晋之时，茶便成为僧道修行中普遍的饮品。盛唐禅宗发展后，几乎"庙庙种茶、无僧不茶"。禅宗清静修行的旨归与茶叶醒神宁心的功用实现了统一。同时，由于僧侣具有一定的学识，使得茶叶的种植和茶文化都获得了长足的进步，故有"名寺出名茶"之说。

　　Buddha meditation has been linked closely with tea. The saying goes. Tea geese aold poetry, the skyis clear and the heart is clear. Tea had become everyday drinking for monks and Taoist priests as early as the Wei-Jin Period. Tea was planted in every temple, and all monks drank tea when Zen was prosperous in the Tang Dynasty. The famous monk Jiao Ran of the Middle Tang

Dynasty has a poetic saying. Zen and tea realize the perfect combination. At the same time, Tea cultivation and culture have progressed considerably through monks' knowledge and skills transfer. Famous tea also comes from those renowned temples. Enjoy tea in the Buddhism place is simple and elegant. Drinking tea in a quiet temple can give unforgettable memories.

亭台楼榭伴荷池
At Various Ancient Pavilions and Lotus Pond

有园林、湖泊和池塘之处，也多有亭台。亭台楼榭、荷花池塘的风景，如果以之佐茶，更加显得妙趣横生。西湖的龙井、太湖洞庭山的碧螺春、洞庭湖的君山银针都是与湖相关的名茶。亭台楼榭在湖光山色中有着增进娱游的独特设计，湖中莲花的香气随风飘摇在游人的四围。荷花素为花中君子，文人雅士多有培植，且制有荷花茶。在亭台休憩处，品茗赏荷放松心情，吟诗作画追慕风雅，身心均得以陶冶舒畅。

There are lots of pavilions in the garden. lakes and ponds The scenery of the park is unparalleled. Drinking tea has various wits in place All famons teas related to. The lakes are West Lake Longjing, Taihu Lake Dongting Green Spiral Tea, and Dongting Lake Jun Mountain Silver Needle Tea. They are designed uniquely. The aroma of the lotus is surrounded. Lotus element is a gentleman among flowers, Many scholars planted the lotus here. People enjoy tea with beautiful scenery to nourish their souls and minds.

茅屋田园家门前
Thatched Cottage in Countryside

戴月荷锄而归，菊花清茶数盏，田园品茶的质朴中充满着诗情画意。茅屋清茶是一种意境美，也是品茶的佳境。数竿潇竹于晚风里摇动，明月挂影镶嵌在闪亮星天，柴门犬吠，人声寂静，田园生活之美。远山奇景中的茶境可以造就高雅脱俗，而茅屋田园的门前小啜，也能享受恬淡怡然的乡野气息。远离喧嚣而不拒绝生活，田园格调也平添了雅趣。

At dawn, I rise and go to weed the field. Then, shouldering the hoe, I walk home with the moon. Simple ceramic cups and their mellow taste can bring you into the fantastic picturesque realm. It is also an excellent place to enjoy tea. The bushes and trees are blowing in the wind. They want the fun of rural life. It can achieve an elegant and refined spirit. People can get away from the noise and hurry of busy working lives. Set a stone table in front of the door. Then chat with friends and enjoy tea soup in a tranquil mood. The beautiful scenery is pleasant, and the wonderful nature is desirable.

竹　林
Bamboo Forest

竹林自古就是饮茶之人所追求的清幽的品茗佳境。置身于翠绿浩瀚的竹的海洋中，倾听着风吹竹叶的天籁之音，品饮一壶清茶，用心去享受这份自然赋予的惬意与宁静，忘却尘世的喧嚣，一切浮躁都销声匿迹，是清雅闲淡的诗情，是恬然优美的画意，更是令人神往的写意。

369

The bamboo forest has been a wonderful place for enjoying tea since ancient times. It shows more brilliant wisdom and the art of living by enjoying the embrace of bamboo. So enjoy the quiet time and delicious tea to escape city noises. It is an elegant and leisure poetry. a tranguil and beautiful painting, and even a captivating freehand brushw-ork.

幽 谷
Secluded Valley

幽谷总透着自然的灵气，清茶总能舒神宁心。踏幽谷，品清茶，是深受古人喜爱的休闲生活。茶香飘逸，泉声绕耳。聆听鸟语清音，山石为景，流水为乐，轻啜一口清茶，别有一番悠然的心境。品茗在于环境更在于心境，在这样一种怡然的氛围中，又怎能不让人心情愉悦。不知这样的意境激发了多少诗人、画家的创作灵感，又有多少文人墨客陶醉于这份茶香悦色的美境中。寄情于山水，茗茶相伴，该是怎样的一种惬意。

The secluded valley expresses the native intelligence and beautiful charm of aura. Surround yourself with a fresh tea aroma. Sit on the bench, listen to the birds' sounds, and sip tea in exclusive surroundings to inspire ideas. Tea tasting is not only about the environment but also about the state of mind. In such a pleasant atmosphere, how can one not feel happy. I don't know how many poets and painters have been inspired by this artistic conception. So many scholars had self-regulation in this beautiful place with tea. How comfortable it is to express oneself in the mountains and rivers, accompanied by tea.

意境之美
The Beauty of Artistic Conception

茶艺篇 Tea Arts

人文环境
Humanistic Environment

古人饮茶多在自然环境中寻找乐趣，现实生活中的人们往往身处闹市而远离山野。人文环境与自然环境相呼应，也可称之为人造环境，在这种环境中，人造景观与文化氛围都是最重要的元素。

The ancients found joy in nature with tea. Modern people live in the downtown center, far away from the mountains and plains. The human environment combines with the natural environment. It can also be called an artificial environment. Artificial landscapes and cultural atmospheres are essential in this environment.

除了建筑与创造的人文特性之外，人文环境还追求一种自然与

人和谐的氛围。茶，正满足了这种需求，能够很好地营造出优雅、淡然的人文气氛。为了享受品茗的乐趣，爱茶之士精心布置、创设了各种人造设施和条件，以求在品茗过程中达到自然与个人深度融合的身心享受。例如，现代茶舍在注重营造返璞归真的自然气息的同时，也倾向于安排脱离城市喧嚣和嘈杂的人文环境要素。书画、音乐、围棋、曲艺均能很好地创造出别致的人文环境。饮茶的环境不拘泥于一定的模式和固有的元素，无论是喧嚣的都市还是僻静的山村，都可以利用已有的环境特点因地制宜地打造出雅致的人文环境。

Besides the architecture and creativity of human creation, the humanistic environment also pursues harmony a tmosphere between nature and human beings. Tea is satisfied with the requirement. It can help create elegant and tranquil human istic atmo sphere.

Tea lovers prepare and create a drinking environment with meticulous care for enjoying tea. It can help achieve a deep amalgamation of nature and human beings. For example, people focus on creating a natural flavor and helping themselves escape the city's hustle and bustle. Calligraphy, painting, music, and Chinese rap can all create a unique humanistic environment. No matter the noisy town and quiet mountain village, they all can make an elegant humanistic climate according to their existing characteristics.

内部环境
Interior Decoration

内部环境是指在人文环境之内富含创意的各种陈设布置，例如馆阁中陈列的名家书法、宁静致远的禅韵、舒缓妙曼的乐韵、古朴天然的茶具与雕饰等。现代茶舍中，同样注意营造品茗时怡然优雅的内部环境。

The interior environment refers to all kinds of creative collocations from the humanities environment, such as the calligraphy, elegant and tranquility Zen rhyme, graceful and leisurely music rhyme, natural and straightforward tea wares and decorations, etc. It also should create an elegant internal environment for a modern tea house.

美妙心境
Wonderful Mood

品茶既要用口也要用心。因此，品茶时候的美妙心境更是不可或缺。美妙心境如其字面意思，即心情要无忧、舒畅、放松，心无挂碍同时能够悠然闲适，不牵缠世俗的烦琐，忘却生活的劳碌。

Tasting tea is not only with the mouth but also with your heart. The excellent mood is indispensable, relaxing, fun, carefree, and leisurely. It can help you keep a distance from everyday customs and forget all the troubles in life.

古人品茶讲究心境的三个层次：首先就是有空闲，即不受琐事牵绊；其次还要内心清静，即淡泊宁静的心情；最后是与三五知音共饮，在品茗过程中伴以闲谈，通过意识的碰撞，达到心灵的共鸣，进而进入启迪性灵、感悟人生的境界。烦恼的时候不妨适度地"将进酒"，悠闲的时候不如细细地品茶。因此，品茶需要有相对美好的心境。

There were three levels of tea drinking. First, you need to drink tea at your leisure. that is not boing hindered by trivial matters Second, we should have a quiet mood with drinking. Finally, you can enjoy tea with some friends. They can explore the resonance of spirit and nature. When you are troubled, it's better to drink in moderation. When you are leisurely, it's better to savor tea carefully. Therefore, you need to taste tea in a beautiful mood.

别致的露天茶座
Fancy Outdoors Teahouse

不一样的环境，自然也会有不一样的心情。露天的茶座总是能平添几分浪漫。夕阳西下的黄昏，欣赏着渐渐退却的红晕，忘却工作的疲惫，只细细品味着茶的芳香，享受着生活的美好。当夜色降临，沐浴在皎洁的月光下，灯光闪烁，繁星点点，茶香四溢。室内别致的装饰也能营造出温馨的露天茶座的感觉，独特的门窗，精巧的花石，再布置出一个唯美的小露台，虽置身室内，却别有一番浪漫典雅的情趣。

People have different moods in different environments. Some prefer to drink tea in the fancy outdoor teahouse for romance. When the sun sets, people can enjoy the beautiful scenery. The whole sky became suffused with bright pink. It can ease the work pressure and soothe the tired heart. When the night comes, bathed in the bright moonlight, lights flashing stars, tea orefflowing. Unigue inteaior decoration can also create a warm open-air teahonse feeling. Sip the tea slowly, then savor the silence. Some other fancy trimmings and unique decorations are permeated with an elegant and modern atmosphere.

图片由 BASAO 提供
Photos by courtesy of BASAO

人声鼎沸的茶馆
Bustling Teahouse

21世纪的中国茶馆蓬勃兴起，使茶文化成为了当今社会不可缺少的精神文化元素之一。生活紧张而忙碌的现代人，已经越来越深地爱上了去茶馆品茶这种调节身心的消遣方式。泡茶馆已成为一种新的时尚。虽然不同地域的茶馆都各具特色，有的诗意儒雅，有的古朴典雅，有的可以边饮茶边听曲看戏，有的则可听书说唱。但无论是哪种风格的茶馆文化，总能吸引爱茶人士的追捧。似乎只有到茶馆中品茶才是真正的嗜茶人。

Chinese teahouses craze makes tea culture become one of the indispensable spiritual and cultural elements in the 21st Century. Those nervous and busy modern people have already enjoyed this pastime deeper and deeper. It has become a new fashion. Even though the teahouse differs in various regions, tea lovers are all cherished. You seem to be a real tea lover when you love tasting tea in the teahouse.

清幽的茶室
Peaceful and Secluded Tearoom

与茶馆相比，茶室更给人一种清幽、宁静的感觉。简单的设计，轻装修重装饰，几张茶几，几把木椅。古香古色的木制门窗，洁净又不失典雅，再配上古朴的紫砂壶，具有中华民族特色的青瓷碗，光是欣赏着这一切就能使人的身心得到放松。斟上一杯清茶，感受着空气中迷漫的茶香，品上一口香茗，这种难得的闲适正是从古至今人们所追求的幽雅和情调。闹中取静，清幽的茶室仿佛喧闹都市中的世外桃源，人们紧张焦虑的情绪在这里得到了一丝缓解。

Compared with the teahouse, the tearoom has a peaceful and secluded feeling. Its design and decoration are simple and elegant. Moderate tea drinking can also help relax and rejuvenate. People sip a cup of tea and enjoy its fragrance inside the room. You can keep quiet in the rustling and bustling city. People can reduce tension and anxiety this way.

雅致装饰
Elegant Interior Decorations

对于品茗而言，茶舍对内部环境的设计和布置，同样体现着对品茗佳境的努力追求。要布置出适宜品茶的内部环境，雅致的装饰品则必不可少。饮茶的环境通常要求清静、洁净，不但舒适还需文化气息浓厚。因此，陈设和布置要能令人感到放松、摒弃杂念、忘却烦恼，从而专注地沉醉于品茗的乐趣中。装饰品多以文化艺术的形式或以茶本身为主题，如名家书法、绘画、匾额题字，古朴典雅的雕刻、摆饰，精致的或独具特色的茶具，以及突出中国传统文化特色的瓷器、刺绣、绳结等艺术品。此外，可以净化空气的观叶植

物也能使人远离尘嚣，亲近自然。

The design and arrangement of the tea house also embody the pursuit of aesthetic modernity. Elegant interior decorations are essential parts of tea culture. It requires a quiet and comfortable place for tea tasting. Therefore, most interior decorations are embodied through philosophy, religion, and the arts. There are famous calligraphies, paintings, horizontal inscribed boards, elegant carvings, and traditional Chinese porcelain wares. Furthermore, an indoor leaf-viewed plant can freshen the air, help people escape the annoying, disturbing public, and be closer to nature.

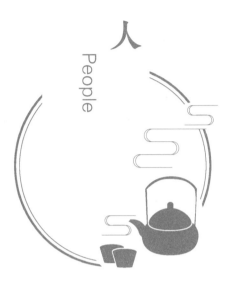

茶艺篇
Tea Arts

　　茶艺本身便是一种茶与艺的结合,通过茶人,将茶中富含的生活、文化以及人生淋漓尽致地表达出来,从而赋予品茗以更强的灵性和美感。茶艺是一种有意味的品茗活动,追求的是人与茶、人与自然以及人与人之间的和谐,是一种精神上的崇高享受。茶艺表演要求茶艺师具有一定的仪表、仪态和仪礼,将个人修养与茶文化的精华相互融合。高水平的茶艺表演能够给人以精神、生活乃至道德观念上的鼓舞、愉悦和导引。

　　Tea arts is a developmental aesthetics with a combination of dynamics and statics. Tea lovers can incisively and vividly express the designers' ideas. It truly represents the practical form of myriad things. People are always in pursuit of harmony between man and nature. As we know, tea art can bring us fantastic enjoyment, delight, and happiness when tired or sad.

　　Tea art specialists must be fastidious about their appearance,

manners, and etiquette when giving a tea show. However, those good tea art performances can create great encouragement and excitement.

茶艺本身有了人的展演,使得自然精神经过人文精神得以再创造,是一种物质和精神上的统一。一个好的茶艺师,要有平和的心态,优雅从容的仪态,以及含蓄、谦逊、诚挚的礼仪动作。茶艺表演要求茶人、茶与艺术浑然于一体,通过固定的技艺和不同的环境,使得品茗过程中的所有内涵与意蕴得到最大的展现。

The people-oriented concept is vital for tea art shows. It is the unity of the substance and mind. A good tea art specialist should have a peaceful mind with noble and elegant manners and implicit behavior. It creates incisively and vividly aesthetic charms. Tea's connotation and implications can be fully presented in different environments.

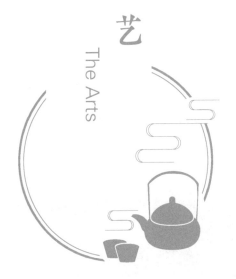

艺 The Arts

茶艺篇 Tea Arts

泡茶的技艺可以说是茶艺活动的主体，是充满诗情画意的艺术活动。茶艺简单的理解即是一种艺术性的泡茶和品茶的学问，是"茶"与"艺"的结合，是泡茶技艺与品茗艺术的结合。茶人在人们日常饮茶习惯的基础之上，根据一定的规则进行艺术加工，向饮茶人和宾客展现冲茶、泡茶、饮茶的技巧。

The skills of tea brewing is the subject of tea art events, full of poetic charms. It also can be thought of as artistic tea brewing and tea tasting. It combines tea and art and perfectly combines tea brewing techniques and tasting art. The hosts show off their brewing and drinking skills.

艺术化的饮茶方式提升了品饮的境界，赋予了茶较强的灵性和美感。茶艺表演中讲求熟练、细腻、规范，还要传达神韵。对茶人的精神内涵、动作手法和语言等都有特定的要求。

Artistic tea drinking can ascend to a new tasting level and provide a visual aesthetic. It gives tea a strong sense of spirituality and beauty. Tea masters should be skilled and exquisite in the tea ceremony. In addition, there may be specific spiritual connotations, movement, and language criteria.

茶艺的精神内涵
Spiritual Connotation of Tea Art

茶艺篇 Tea Arts

 崇静尚简 Calm and Simplicity

宁静简素之美是中国茶艺的特点之一。静，既指茶人内心的清静、淡泊，也指环境的宁静澄明。简，既指茶人的清丽简约，也指陈设的素淡不繁复。

The beauty of tranquility and simplicity is one of the characteristics of Chinese tea art. Quiet refers to the inner quiet and indifferent of tea people and also refers to the tranquillty and clarity of the enviro ment. Simplicity lefers the beanty and simplicity of tea people, and also refers to the plain and not complicated furnishings.

崇静尚简的境界是对心境与环境、人境与物境的多维要求，是人与自然的和谐统一。

It is the harmony between man and nature and can promote the inherent mental state. The cultural atmosphere and aesthetics are closely related.

内省求真
Introspection and Pursuing Truth

茶艺内省求真的特点具有哲思的禅意和美感。将禅所追求的内省和顿悟融入茶艺之中,增加了品茗时内心的宁静超然之感;将"求真"引入茶艺之中,更增强了对尘世琐事的超脱及豁达开朗的心灵境界。

The tea ceremony has the characteristice of introspection and desire of truth contain zen buddhism sense and aesthetic feeling. Integrating introspecgior and epiphany, Wich zen Buddhism is seeking, into the tea ceremony, can bring us inner peace with a detached attitude when sipping tea. Introducing introspection into the tea ceremony can further inprove the detachment facing trifles and help reach the realm of open-minded and happiness. It is also an inherent introspection and epiphany and can make us feel quiet with a detached and free attitude. Introspection can help you become calm and indifferent. First, it represents an attitude of modesty, then brings refreshing harmony of nature.

茶艺篇

Tea Arts

茶艺的种类
Type of Tea Arts

茶艺是生活与艺术的结合，是雅致的生活方式，是有意味的精神享受。狭义的茶艺是指冲泡技艺和饮茶的艺术，而冲泡技艺和饮茶正是来自于人们的日常生活习惯。

Tea art combines life and art, representing an elegant lifestyle, spiritual pursuit, Tea art is about the brewing and drinking technique and a life attitude.

中国的茶叶品种繁多，地域特色明显，茶艺也不拘一格。茶艺以人、茶叶主体、冲泡方式和茶具为标准，可以分为不同的种类：按历史可以分为传统和现代；按地域可区分为南、北派及港、台；按用途可分为表演型和待客型；按主体类型可分为高雅、流行以及皇室、贵族、宗教、文士、平民等。虽然标准有不同，茶艺的类别也千秋各异，但其内在神韵却具有高度的统一性。

China has a variety of famous tea with distinct geographical characteristics. Tea ceremonies do not stick to one pattern. Tea art can be divided into different types based on the creterias of people、tea、brewing methods and tea set. According to another historical period, it takes people and tea with those tea wares to divide into different types, traditional and modern. It also can be divided into southern, northern, and Hong Kong and Taiwan types. on the basis of its users, it has performance and guest types. The style type can be divided into elegant, famous, royalty, aristocracy, religion, literati, civilians, folk customs, etc. Even though there are diverse standards and categories of Tea Ait, they all have high consistenly on inner natural beauty and charm.

分类的方法
Classification Method

按茶事功能划分
Divided by Function

按茶事功能划分，茶艺大致分为表演型和待客型。表演型茶艺是由一个或几个茶艺表演者演示茶艺技巧，重在观赏。大多数观众并不能鉴赏到茶的色、香、味、形，更品不到茶韵。然而，表演型茶艺在推广茶文化，普及和提高泡茶技艺等方面都有良好的作用。

Tea arts ceremonies can be divided into performative and entertaining types. one or severe tea art performers demonstrat can present tea art skills for appreciation. But most guests cannot appreciate its color, fragrance, taste, and appearance, let alene Yun of tea. However, a performative tea ceremony can promote tea culture and increase its popularity. It also can improve brewing skills and tasting skills.

按茶叶种类划分
Divided by Tea Variety

因为中国茶叶种类各有差别，适宜的茶具、冲泡的水温和方式方法等也各有不同。因此，按照茶叶的种类来划分，茶艺的类型主要有乌龙茶茶艺、绿茶茶艺、功夫红茶茶艺、花茶茶艺、普洱茶茶艺等。

China has a variety of teas. Some tea brewing can vary widely, such as the brewing utensils, brewing temperature, and brewing methods. According to tea variety, tea art ceremonies mainly conclude the Oolong Tea ceremony, Green Tea ceremony, Gong Fu Black Tea ceremony, Scented Tea ceremony, Pu Er Tea ceremony, etc.

按饮茶器具划分
Divided by Drinking Utensils

在泡茶茶艺中，因使用泡茶茶具的不同可以分为壶泡法和杯泡法两类。壶泡法是指在茶壶中泡茶，然后将茶分别倒入小杯中再饮用的方法；杯泡法则是直接在茶杯或茶盏中泡饮。杯泡法茶艺还可分为带盖杯泡法和玻璃杯泡法两个小类。

It has pot brewing and cup brewing methods in the tea ceremony. Pot brewing is how people use a teapot to brew tea leaves, then pour tea soup into several cups for tasting. The Cup brewing method is similar. People directly brew tea in the cup. The Cup brewing method can be put into two categories of cup brewing and glass cup brewing.

按饮茶主体划分
Divided by Drinking Style

以人为主体的茶艺可分为宫廷茶艺、文士茶艺、宗教茶艺等。宫廷茶艺采用模仿宫廷茶宴的形式，具有场面宏大、礼仪繁琐、气氛祥和、讲究技巧和所用茶具力求奢华的特点。

It can also be divided into the palace tea ceremony, literati tea ceremony and religion tea ceremony, etc. The Palace tea ceremony is an activity that imitates an environment with qrand slene tedious etiquette and a relaxed atmosphere. The skill and tea types are lavish.

按民俗划分
Divided by the Culture and Folk Customs

中国民族众多，在长期的饮茶历史中，很多民族发展并世代相传着自己特有的品茶习俗。如藏族的酥油茶、蒙古族的奶茶、维吾尔族的香茶、回族的罐罐茶、土家族的擂茶、白族的三道茶、苗族的油茶、畲族的宝塔茶、布朗族的酸茶、傣族和拉祜族的竹筒茶以及纳西族的"龙虎斗"茶等。泡茶原料、冲泡形式和礼法、寓意等也各有不同。多姿多彩的民俗茶艺，使得中国茶文化更加丰富绚丽，闪烁着各民族的独特光芒。

There are many transnational ethnic groups in China. During a long drinking history, most minority areas have developed and transmitted their tea-drinking customs from generation to generation. There are Tibetan buttered tea, Mongolian milk tea, Hui minority Guan Guan Tea, Miao Minority Lei Tea, She Minority Baota tea, Blang Minority Sour Tea, Dai and Lahu Minorities Scented Tea in the bamboo box, and Naxi minority Dragon Fighting against a Tiger tea, efc. These types have different tea types, suchas brewing methods, etiquette, and morals. Those colorful folk customs make Chinese tea culture more gorgeous. Moreover, it brings a brighter future for their own tea culture.

以冲泡方式划分
Divided by Brewing Methods

以冲泡方式作为茶艺分类标准，相对较为科学。在中国的饮茶史上，曾经出现煮、煎、点、泡四类饮茶法。而由此四类推衍形成的茶艺有煎茶法、点茶法和泡茶法三大类。遗憾的是，发展流传至今的只有在明朝中期形成的泡茶茶艺，其他几种却未能传承下来。

It is more scientific to classify tea art ceremony types with brewing methods. Four brewing methods have appeared in Chinese tea-drinking

history. Which contains boiling, steaming, whisking, and brewing. Then Steam Tea Method, Dian Cha Method (whisk method), and Brewing Tea Method are established. Unfortunately, only the Mid-Ming Dynasty tea ceremony can be passed so far. Other tea ceremonies cannot be passed on.

按时期划分
Divided by Period

中国茶艺经历了一段长期的发展过程，在不断积累经验总结技艺的基础上，才形成了色彩纷呈的茶艺文化。传说中神农氏口尝茶叶的方法，魏晋时期的配茶，唐朝时期的烹茶，宋代时期的斗茶，明清之际的饮茶，显示了茶艺发展的历史轨迹。

The Chinese tea ceremony has experienced long periods of developing history. It accumulates experience in practice to create a colorful tea culture. According to the tale, the Shen Nong tasting method, the blending method in the Wei and Jin dynasties, the brewing method in the Tang Dynasty, the tea competition in the Song Dynasty, and the drinking method in the Ming and Qing Dynasties can show the historical trace of Tea Ait's evolution.

当代的茶艺大多在吸收不同时期茶艺技巧和精神的前提下，依照自身地域特色和茶叶品质进行了创新性的精进，从而使得茶艺本身便有着浓重的历史气韵。

The modern tea ceremony absorbs colorful skills and spirits from different periods. After the innovative development, it has a solid Chinese history aura with its great regional feature and tea quality.

茶艺篇

茶的鉴别
Identification of Tea

"茶者，南方之嘉木也。""嘉"者，泛指一切美好的事物，如何鉴别茶的优劣好坏，从中选出佳品，也是一门深奥的学问，需要从茶的产地、外形、色、香、味以及冲泡之后的茶汤滋味等各方面综合比较鉴别，以选出真茶、好茶。

由于茶的类型和特性的多样性，给所有茶叶制订一套统一的评判标准并不现实。不同品性的茶有不同的鉴别要求，而不同人的口味千差万别，泡饮习惯更不相同，应依据个人的喜好、身体状况等选择适合自己的茶叶，利用好水佳器，将现有的茶叶泡出最佳的风味，才是饮茶之佳境，才能从中得到最高的品茗乐趣。

Tea tree is a premium timber in south China. It can be referred to as all the glorious things. Identification and distinction of tea is a profound technology. You must identify tea leaves in the following aspects: a place of origin, shape, appearance, color, aroma, taste, etc. Due to its diversity in its types and characteristics. Different kinds of tea should be evaluated with varying criteria of identification. And people's tastes vary widely. So we should select the tea depending on our taste and preference. Brew out its best flavor with good-quality water and brewing utensils. You can get the supreme kind of pleasure from drinking tea.

茶艺篇

Tea Arts

茶的鉴赏
Appreciating Tea

观察外形
Observing Tea Shape Appearance

鉴别茶叶时，外形是第一直观要素，对成品茶外形的审查标准是干燥、匀整、鲜嫩、干净。品质优良的茶叶，任意拈一根茶条，可轻易折断并碾成粉末，即表明干燥程度达标。

Tea shape appearance is one of the most intuitive visual elements of appreciation. The specific standard of appearance shall be dry, evenly fresh, tender, and neat. You can easily break and roll the excellent quality tea leaves into powders. It shows the standard dryness.

无论是圆状、扁状或条状的散茶，优质茶的形状、大小均会保持一致：圆状的粒粒大小相同；扁状的条条厚薄相等；条状的长短匀称。这些都表明制茶技术高，采摘规格严，鲜叶的采摘一芽几叶全部相同，工艺精良。

The size and shape should be kept the same for one type of the excellent quality tea with the round, flat, or strip loose tea leaves. The round shape is the same size, the flat shape is the same thickness, and the strip length is uniform It shows the advanced processing technique and strict plucking standard. All plucked tea leaves are almost the same.

好的茶叶条索紧结，在外形上卷曲优美、完整匀称，表面上油润、洁净，没有断梗碎枝、木质纤维、粉末及杂物，而且芽毫多、叶质嫩；反之，则茶叶粗糙、叶质老、条索粗松、色泽花杂、身骨轻、净度差。

Good quality tea leaves are bold and tight in shape. It is curly and elegant. Whole tea leaves are even and neat without broken stems, powders, and foreign matter. It also has plenty of tips and fuzzes. Conversely, the lousy tea leaves are coarse, hard, and mixed with a harsh odor. Moreover, they are dull and combined with worse neatness.

辨别颜色
Distinguishing Tea Color

茶叶色泽完美呈现需将茶树品种、采摘的茶园条件、叶色等多方面因素进行配合，再配以精良的加工，才能使成茶色泽匀正，且符合该类茶叶所特有的深浅、润枯和鲜暗标准。干茶叶的色度依茶类不同有很大差异。以绿茶为例，绿茶以茶芽鲜嫩为佳，呈黄绿色，如果颜色浅绿，则低一等级，而颜色深绿，则该种茶叶为最差。鲜嫩茶芽所制成的茶叶，应色泽油润，并隐含着光泽。

The uniform color of tea depends on many factors. It needs cooperation in many aspects, such as tea variety, garden condition, tea leaves origin color, and excellent processing. With these factors in order to make the xolor of the tea uniform, uhile in line with the chara cterist ics of this tea depth moist and clark. And the original dryness, humidity, and brightness are also significant. Due to many tea varieties, the chrominance of dry tea is also quite different. Take Green Tea as an example. Those tender and yellowish-green color tea buds are the best. Tea leaves with pale green color are inferior. The deep green leaves are the worst. Those tea leaves processed with fresh tea buds have a bloom color. But tea leaves would be dull blueish green with too tender buds as material.

但是如果采摘不适，过嫩则茶叶青绿，颜色发暗；过老则颜色干枯，不生动。当然，茶叶的色度和制作工艺也有很大的关系，如果杀青不充分，也会造成茶叶光泽不足，不整齐。而制作工艺粗劣，即使鲜嫩的茶芽也会变得粗老、枯暗。

Conversely, tea will be dry and dull. But, of course, its color also depends on the processing technique. Tea leaves will be dull and uneven, with not enough fixation. The fresh tea buds will also be aged and dull, with worse processing techniques.

掌中感受
Feeling in the Palm

感官的直观感受是评茶时的一个重要标准。取适量的干茶叶置于手掌中，通过感受掌中的重量感，可以判断茶叶的精粗度和紧密度。还可用手指将掌中的茶摊开，减少重叠，以利于对单个茶叶进行检查。抓茶时，要注意茶叶的光滑紧凑度，看是否有茶末、茎梗、茶片等杂质。

The intuitive sense is one of the primary standards of sensory tea tests. Take several tea leaves into the palm of your hand to feel their weight. It can estimate its coarse degree and tight degree. You can also spread tea leaves in the palm to check the available tea leaves. When grabbing tea, pay a ttention to the smoothness and compactness of the tea to see if there are impurities such tea dust, stems, and flat tea. You should select tea dust, stem, and flat tea leaves when you take tea leaves.

闻香气
Aroma Smelling

香气是茶叶的灵魂。各种茶都有各自不同的茶香。选茶时，干茶的香气高低、强弱、持久程度、纯正与否都是重要的评判标准。双手捧茶叶，靠近鼻端轻嗅，一是辨别香气的高低，二是嗅闻香气的纯正程度，以高锐、鲜爽、浓烈、持久、无异味者为佳。凡高香茶、新茶、足火茶，干嗅时香气高、气味正；而劣质茶、陈茶或水分含量多的茶，干嗅则香气寡淡、低沉，且夹杂异味。

The aroma is the soul of tea. Each kind of tea has a unique scent. Its aroma amount and aroma quality are the essential indicators to measure the quality of tea. When you hold the tea in your hands, Smell tea leaves with your nose close to distinguish and experience their purity. That tea leaves with high, sharp, fresh, brisk, strong, pure, long-time lasting, and without peculiar smell. Invevsely, The aroma is thin, plain, low, and dull of lower quality excessive water content and worse aged tea leaves.

观汤色
Observe Color of Tea Soup

鉴别过干茶后，还必须开汤审评，才能进一步判断茶叶优劣。通常取茶叶3~5克，用150~200毫升沸水冲泡，静置3~5分钟后，将杯中茶汤倾入另一空碗中，审看茶汤的色度、亮度及清浊程度。好的茶叶茶汤香气浓郁纯正，闻之神清气爽，汤色显浅绿或呈黄绿色，清而不浊，明净澄澈；次质茶叶则亮度差，无光泽，透明度低，暗淡混浊。

After observing the tea's appearance, we must brew tea soup for sensory tests. It can help us to distinguish and evaluate tea leaves. We usually take three to five grams of tea leaves to brew with one hundred and fifty to two hundred milliliters of boiled water. Then stew it for three to five minutes. Then pour tea soup into another empty bowl. We can observe its Chroma, brightness, and turbidity. The sound-quality tea leaves have a strong and pure aroma to refresh you. The color of tea soup is tender green and yellowish-green. It is pure, clear, bright, and not muddy. The lower degree of tea leaves also has lower quality and brightness with low transparency. In addition, they are dull with lots of suspensions.

品茶味
Appreciating Tea Taste

茶汤滋味根据个人口味及茶叶种类不同，也存在各式各样的口感，鉴别标准并不统一。一般来讲，绿茶茶汤鲜爽醇厚，初尝略涩，后转为甘甜；红茶茶汤甜味更浓，回味无穷；花茶茶汤滋味清爽甘甜，鲜花香气明显。而所有类型茶叶的劣质茶滋味都很平淡、苦涩，且略有焦味，无回味或者回味短。

The taste of tea soup should vary according to personal taste and tea variety. The distinguish standard is not the same. The Green Tea soup is brisk and mellow. You will feel a bit of astringency at first. Then the taster would be sweet after then. Black Tea is a sweet and strong flavor long after taste. It can tead a person to endless of tertases. Scented Tea is brisk and sweet with a strong floral fragrance. The lower degree of all other tea types is plain, thin, and bitter, slightly burnt. Its aftertaste is weak.

看叶底
Appreciating Brewed Leaves

茶叶的叶底往往最能反映茶叶的真实面目。叶底完整，那么茶叶的采摘和加工就会精细。叶底明亮有光泽，那么泡出的茶水也会明亮有光泽。叶底柔软，说明原料鲜叶比较细嫩。相反，叶底粗硬说明原料鲜叶粗老。好的叶底也能够给人带来赏心悦目的观感。

The brewed leaves have been one of the most visible aspects of appreciating tea leaves. The whole tea leaf shows the fine plucking and processing technique. Tea soup would be glossy with those bright material tea leaves. Those brewed tea leaves can show that their material leaves are thin and tender. Conversely, material tea leaves are coarse and aged. Those lovely brewed leaves also are the Feast for the eyes.

外部环境
Appreciating Environment

评茶应在专门的评茶室中进行，评茶室要求安静、清洁、干燥、空气新鲜、无异味、光线明亮柔和，避免阳光直射。评茶室北面开窗，设置黑色斜突30°的遮光板，以统一光源的均匀。室内墙壁要刷以白色，并引进自来水，设置洗盘池和出水导管等。

Tea sensory tests should be held in the specific testing room. The testing room should be quiet, clean, and dry in the fresh air without a peculiar smell. Light is soft and even. And space should be kept out of direct sunlight. The windows should be set north with a black shading plate for even lights. The room should be painted white. It also has a pool for washing and outlet conduit, and so on.

茶饮篇

Tea Drinking

茶之为饮,发乎神农氏,闻于鲁周公。现如今,茶作为一种饮料在世界范围内得到了广泛的推广。

Tea originated from the Shennong and originated from Duke Zhou of Lu. Nowadays, Tea gets extensive development worldwide as a beverage.

茶经六之饮
Chapter 6: Tea Drinking

茶饮篇
Tea Drinking

翼而飞，毛而走，呿而言，此三者俱生于天地间，饮啄以活，饮之时义远矣哉！至若救渴，饮之以浆；蠲忧忿，饮之以酒；荡昏寐，饮之以茶。

Birds have wings and fly, beasts have abundant fur and run, and people can speak. These three are all born in heaven and earth. They are relying on drinking water and eating to sustain life activities. It can be seen that drinking has a significant and far-reaching effect. To quench thirst, drink water; To relieve depression and excitement, one should drink alcohol; To refresh oneself and relieve drowsiness, one should drink tea.

茶与健康
Tea and Health

　　茶叶中含有丰富的营养元素，因此饮茶也有着多方面的保健功效。茶作为人们日常生活中健康的饮品之一，有着咖啡等其他饮料等所不具备的诸多优点。同时，饮茶也有着多样的禁忌或不宜，不当的饮茶方法和习惯对人体的健康也会造成损害，不过科学合理的饮茶方式对人们的健康和心情却能起到很大的助益。总之，因为茶对于人身体健康无比重要，所以在很多地区也便有了"宁可三日无饭，不可一日无茶"的说法。

　　Tea contains a rich nutritional composition. Drinking tea has a powerful effect on health care, and as one of the healthiest beverages in our daily oife, tea with significant advantages over coffee and drinking. There are also some restrictions and taboos on drinking tea. Incorrect drinking methods and habits could even damage health. People can benefit much from the scientific and proper drinking method. There is a saying in many areas. It is better to be deprived of rice for three days than starved of tea for a single day.

茶的健康元素
Tea Health Element

茶饮篇 Tea Drinking

咖啡因
Caffeine

咖啡因是茶和咖啡等饮料中的活性成分，是一种生物碱。在作为药品时可用于心脏和呼吸促进中枢神经兴奋剂。对人体有以下作用：振奋精神、提高思维能力、预防疲劳、提高工作效率、减少支气管痉挛、治疗咳嗽、化痰，等等。咖啡因也可以作为治疗心肌梗死的辅助药物。咖啡因有助于利尿和调节体温。

Caffeine is the active constituent of drinks such as tea and coffee, an alkaloid. It can be used as a Cardiac and respiratory stimulant in applying for medicine. The following effects on the human body: raise the spirit, promote thinking ability, prevent fatigue, improve working efficiency, reduce bronchospasm, cure cough, and reduce phlegm.

Caffeine also can serve as an adjuvant drug for myocardial infarction. In addition, caffeine helps to promote diuresis and regulate your temperature.

茶多酚类物质
Tea Polyphenols Substance

茶多酚是儿茶素、类黄酮、酚酸和花青素的化合物。茶与其他植物的主要区别在于富含多酚类物质。绿茶中的茶多酚约占总量的15%～35%。红茶中的茶多酚由于发酵程度较高而被氧化，所以它的含量是10%～20%。

茶多酚可以帮助降低血糖和血脂、促进血液循环、抑制动脉粥样硬化和抗氧化，还可以延缓衰老、消炎、抗病毒、抑制癌细胞增殖、消除口臭等。此外，茶多酚还可以保护人的大脑。它有助于防止电磁辐射，被誉为电脑时代最好的饮品。

Tea polyphenols contain catechins, flavonoids, phenolic acids, and anthocyanin. Tea leaves can be preserved with polyphenols for a long time. The main difference between tea and other plant is abundant polyphenols. The tea polyphenols of Green Tea are about 15% to 35% of the total amount. Due to its higher fermentation degree, tea polyphenols of Black Tea have been oxidized. So its content is 10% to 20%.

Tea polyphenols can help lower blood glucose and serum lipids, promote blood circulation, inhibit atherosclerosis and antioxidants, retard the aging process, diminish inflammation, fight the virus, inhibit cancer proliferation cells, dispel halitosis, and so on. Furthermore, tea polyphenols can also protect your brain. It can help to prevent electromagnetic radiation. It was reputed as the best drink in the computer age.

维生素是人体维持正常代谢所必需的六大营养要素（糖、脂肪、蛋白质、盐类、维生素和水）之一，在茶叶中的含量也十分丰富，尤其是维生素B、C、E、K的含量。维生素B可以增进食欲；维生素C可以杀菌解毒，增加机体的抵抗力；维生素E可抗氧化，具有一定抗衰老的功效；维生素K可以增加肠道蠕动和分泌功能。因生理、职业、体质、健康等各方面的情况不同，人体对各种维生素的需要量也各有差异。通过饮茶摄取人体必需的维生素，是一种简易便捷的健康方式。

Vitamin is one of six main nutrition elements to maintain normal metabolism. The rest five are sugar, fat, protein, salt, and water. Tea is rich in vitamins, especially Vitamin B and CEK. Vitamin B can increase appetite. Vitamin C can help to sterilize and detoxify to better your immune function. Vitamin E has antioxidant and anti-aging benefits. Vitamin K can increase intestinal peristalsis and enhance secretory function. Other individuals need different requirements because of the difference in physiology, occupation, habitus and health. It is an easy, convenient, healthy way to consume vitamins by drinking tea.

矿物质
Mineral Substance

矿物质又称无机盐，它是人体内无机物的总称。和维生素一样，矿物质是人体必需的重要元素。钾、钙、镁、锰等11种矿物质在茶中含量丰富。矿物质主要适合与酶结合，促进代谢。如果人体内矿物质不足就会出现许多不良症状，比如钙、磷、锰、铜缺乏，可能引起骨骼疏松；镁缺乏，可能引起肌肉疼痛；缺铁会出现贫血；缺钠、碘、磷会引起疲劳，等等。因为茶叶中矿物质含量的丰富，多饮茶可以促进新陈代谢，保持身体健康。

The mineral substance is also called inorganic salt. It is a general term for inorganic substances in the human body. The mineral substance is also one of the essential elements necessary for human health. Tea contains eleven minerals, such as potassium, calcium, magnesium, manganese, etc. The combination of mineral substances and enzymes can improve metabolism. Some symptoms exist if the inorganic salt is insufficient. For example, lacking calcium, phosphorus, manganese, and copper will cause skeletal damage and osteoporosis. If magnesium deficiency is severe, there will be some symptoms of paresthesia, muscle spasm, and tantrum. Iron can affect the thesis of Hemoglobin. It is easy to be fatigued if you lack Sodium, iodine, and phosphorus. There are abundcmt the mineral substance in tea leaves. So frequently, drinking tea can promote metabolism and improve general health.

氨基酸是一种分子中有羧基和氨基的有机物，它是人体的基本构成单位，与生物的生命活动密切相关，不仅是生命的物质基础，也是人体进行代谢的基础。

在茶中含有氨基酸约 28 种，例如蛋氨酸、茶氨酸、苏氨酸、亮氨酸等。这些氨基酸对于人体机能的运行发挥着重要作用，例如亮氨酸有促进细胞再生并加速伤口愈合的功效；苏氨酸、赖氨酸、组氨酸等对于人体正常的生长发育并促进钙和铁的吸收至关重要；蛋氨酸可以促进脂肪代谢，防止动脉硬化；茶氨酸有扩张血管，松弛气管的功效。茶中含有的氨基酸为人体生命的正常活动提供了必需的要素。

The amino acid is a kind of organic compound with carboxyl and amino groups. It is the basic unit of human beings and is closely related to vital movement, the material basis of life, and many metabolic activities.

About twenty-eight amino acids are inside tea leaves, such as methionine, theanine, threonine, leucine, etc. These amino acids play an essential role in the functioning of the human body. Such as leucine can promote cell regeneration comprehensively and accelerate the healing of wounds. Threonine, lysine, and histidine can help promote growth and absorb calcium and iron. Methionine can help to accelerate and promote fat metabolism to prevent arteriosclerosis. Theanine has the dilation of blood vessels and tracheal function. The amino acid in tea is an essential nutrient element for human beings.

蛋白质
Protein

蛋白质对人类的生命至关重要。蛋白质的基本组成物质便是氨基酸。人的生长、发育、运动、生殖等一切活动都离不开蛋白质，可以说，没有蛋白质就没有生命。人体内蛋白质的种类繁多，而且功能也各异，约占人体重量的16.3%。茶叶中蛋白质的含量占茶中干物的20%～30%，其中的水溶性蛋白质是形成茶汤滋味的主要成分之一。

Protein is essential for human beings. The primary component of protein is an amino acid. Protein participates in the conformation of cells and the activities of the organism. Therefore there is no life without protein. There is a great variety of proteins in the human body. They have different functions and about 16.3% of the total weight. The protein content in the tea leaf is about 20% to 30%. Water-soluble protein is one of the main components to create the taste of tea soup.

糖类化合物
Carbohydrates

糖类是自然界中普遍存在的多羟基醛、多羟基酮以及能水解而生成多羟基醛或多羟基酮的有机化合物。糖类化合物是人体所需能量的主要来源。茶叶中的糖类有蔗糖、淀粉、果胶、多缩葡萄糖和己糖等。由于茶叶中的糖类多是不溶于水的，所以茶的热量并不高，属于低热量饮料。茶叶中的糖类对于人体生理活性的保持和增强具有显著的功效。

Carbohydrates are polyhydroxy aldehydes and ketones or substances that hydrolyze to yield polyhydroxy aldehydes and ketones. They are the body's main source of energy. There are several sugars, such as sucrose, starch, pectin, glucose, hexose, etc. Due to most sugars in tea being insoluble in water, so their heat is not high. and tea is a kind of low-calorie drink. Carbohydrates in tea can help human to maintain and improve physiological functions significantly.

芳香物质
Aromatic Compound

芳香物质是具有挥发性物质的总称，茶叶中的香气便是由这些芳香物质形成的。但是在茶叶成分的总量中，芳香物质并不多，只占到0.01% ~ 0.03%，虽然含量不多，但种类却非常丰富。茶叶中的芳香物质主要由醇、酚、醛、酮、酸、酯、内酯类、含氮化合物、含硫化合物、碳氢化合物、氧化物等构成。因为不同品类的茶叶中成分含量的差异，所以茶叶会有不同的香气。而芳香物质不仅能使人神清气爽，还能够增强人体生理机能。

The aromatic compound is a general term for all volatile substances. These aromatic compounds create a tea aroma. But there are not many aromatic substances its proportion is only 0.01% to 0.03% of the total amount. Even though the entire content of aromatic substances in tea is low, there are abundant types of tea. Tea aromatic compound comprises alcohols, phenols, aldehydes, ketones, acids, esters, lactones, nitrogenous compounds, sulfur compounds, hydrocarbon, oxidizing material, etc. Because of the differente in the compositon content of different types of tea, tea will make different fragrance. Aromatic compounds hot only can difresh your mood but also enhance your physiological function.

其他成分
Other Components

茶叶中的有机酸、色素、脂类、酶类和无机化合物以及上述成分与健康有关。其有机酸和酶能促进新陈代谢，增强免疫力。脂类物质可以起到调节渗透压的作用。

茶叶营养丰富，具有良好的保健作用。

The organic acids, pigments, lipids, enzymes, and inorganic compounds in tea leaves and the above components are related to health. Its organic acids and enzymes can enhance metabolism and strengthen immunity. In addition, lipid granules can play a role of regulating osmosis. Tea has a good health care effect with its rich of nutrients.

茶饮篇 Tea Drinking

茶的保健功效
Health Effect of Tea

安神醒脑
Calm the Nerves and Restore Consciousness

茶叶中含有的咖啡因可以刺激大脑感觉中枢，使其更加敏锐和兴奋，从而起到安神醒脑、解除疲劳的作用。在感觉到心身倦怠的时候，泡上一杯清茶，闻着缕缕的清香，品饮着茶汤的舒爽，精神自然会慢慢饱满起来，已有的困倦和劳累也会得到很好的缓解，从而使思维清晰，反应敏捷。这便是茶带来的安神醒脑的良好功效。

Caffeine is found in tea and can light up the brain's sensorium. It can make the sensorium more excited and acute for refreshing play a rolein calming and waking and relieving fatigue. When feeling mentally and physially tired, make a wp of tea, smell the frogrance of the fragrance, and drink the comfort of the tea, the spirit will raturally be slouly full up, and the existry sleeping and fallgue will be well alleviated. so as to thinking clenrly and respond guiokly. This is the good effect of tea to calm the mind.

防龋固齿
Prevent Tooth Decay

茶具有防龋固齿的功效和它本身含有的健康元素有关。首先是氟元素，茶中含有较多量的氟元素，而适量的氟是抑制龋齿发生的重要元素。因此，一些牙膏中也以添加氟元素的方式来达到更好的防蛀效果。

Tea can help to prevent tooth decay. This is because tea contains a much more Fluorine element. It is the most crucial element for restricting bacteria growth that promotes cavities and tooth decays. Therefore, we also add some Fluorine elements in some toothpaste.

消渴解暑
Relieve Thirst and Summer Heat

作为一种健康饮品，茶可以解渴。夏天喝茶可以使人凉爽。这是一种很好的夏季饮料。在花园里喝茶还是一种美妙的体验。

As a kind of healthy drink, tea can relieve thirst. We suffered a lot in the hot summer. Drinking tea can cool people in the summer. It is an excellent summer drink. Enjoying tea in the garden is a beautiful experience.

BASAO 冷萃茶
BASAO Cold Brew

在恒定低温的萃取条件中，通过长时间浸泡让茶叶缓慢释放滋味，茶汤可以更完好地保留氨基酸、维生素 C 等有益物质。还能降低咖啡因等元素的析出量，因此茶汤清凉甘甜，是适合于对咖啡因敏感、肠胃不适的饮茶者的新兴健康饮茶法。BASAO 精选袋泡茶中最适合冷萃制作的五款，搭配简易上手的冷萃指引说明，开创更科学的"断舍离"饮茶道。

Tea leaves slowly release flavor through a long immersion in stable low-temperature extraction conditions. Tea soup better retains amino acids, vitamin C, and other beneficial substances. Reduce the amount of caffeine and other elements precipitation, so the tea soup is cool and sweet, suitable for caffeine sensitive, gastrointestinal discomfort of the drinkers of the new health tea method. BASAO selects five tea bags that are most suitable for cold extraction and creates a more scientific way of "simplifying life" tea drinking with easy and hands-on instructions for cold extraction.

清新口气
Fresh Breath

人们在用餐之后往往会有一些残余物遗留或者黏附在牙齿的表层或者牙缝中，长时间积聚之后再经过口腔细菌的发酵作用，就会出现异味。饮茶可以起到很好的清新口气的作用。这主要是因为茶中茶多酚类化合物对存在于口腔中的菌类有很好的预防和杀灭效果，同时茶皂甙的表面活性作用也可以起到清除口臭、清洁口腔的作用。

People often have some remnants left or attolhed to the surface of the teeth or between the teeth There is always some food debris in the teeth after dinner. And after a long time of a ccumula tion, It will ferment with oral bacteria to cause bad breath. Drinking tea can help you get fresh breath. Tea polyphenols compound can prevent and kill fungus bacterium. And surface-activated action of tea saponins can also play a role in removing bad breath and cleaning the mouth.

解毒醒酒
Dispel the Effects of Alcohol

说起饮酒对于肝脏的伤害，大家并不陌生，而饮茶可以帮助解毒醒酒，这点也是众所周知。这主要是因为茶中含有的大量的维生素C和咖啡因的作用。维生素C可以促进酒精的代谢，使得肝脏的解毒功能增强。其次，咖啡因的提神作用可以使昏沉的酒醉头脑变得相对清醒，同时还能缓解头疼并促进身体代谢。因此，酒醉后适量饮茶，能达到很好的解毒醒酒的效果。

Drinking too much alcohol is no stranger to the damage to terrible for your liver. And tea can help dispel the effects of alcohol is also well known. It is principal because Vitamin C and Caffeine are essential in relieving alcohol toxicity. It can promote the metabolism of alcohol in the liver and enhance the detoxificotiun of the liver. And Caffeine can refresh your mind and at the same time relieve headoches and boost meta bolism. Therefore, drinking tea after drinking alcoholhas a good detoxification effect.

排毒美颜
Detoxing and Beautifying

经常饮茶可以有效清除体内重金属所造成的毒害作用。研究证明，在人们日常生活中有一些重金属如铜、铅、汞、镉、铬等，通过饮食、空气等方式进入到身体之中，从而对人体造成很大的损害。茶叶中的茶多酚类化合物可以对重金属起到很好的吸附作用，能够促进重金属在身体中沉淀并被排出。

Regular tea drinking coon effeccively remove the toxic effert caused by heaivy metals in the booly. Tea also has a healthy adsorption function. It has been proven that some heavy metals such as copper, lead, mercury, cadmium, and chromium can be cumulated in organisms via skin exposure and food intake to bring apparent toxic effects. Tea polyphenols compounds can absorb heavy metals effectively and discharge metabolic waste.

消食去滞
Promote Good Digestion

人们在酒足饭饱之后往往会出现口渴和食物淤滞的感觉，而这时候饮茶就是最好的选择，可以起到消食去滞的效果。因为茶叶中咖啡因和黄烷醇类化合物的存在，使得消化道的蠕动能力增强，促进了食物的消化。同时饮茶也预防了消化器官炎症的发生，这是因为茶多酚类化合物会在消化器官的伤口处形成一层薄膜，从而起到保护作用。

Drinking tea also can help to promote good digestion. The caffeine and Flavanone compounds can enhance the alimentary canal's peristalsis to promote the faster transit of food through the gut. Tea can also protect your digestive organ with its tea polyphenols compounds.

增强免疫力
Enhance Immunity

个人的免疫力固然跟自己本身的体质有关，但是通过适当科学的方法也可以增强自己的免疫力。饮茶就是一种便捷又健康有效的方式，因为茶中含有的健康元素可以有效地抵抗细菌、病毒和真菌。茶叶中含有较高量的维生素C，可以有效提高免疫力。同时，也有研究认为茶叶里含有的氨基酸也能增强身体的抵抗力。总之，饮茶对于身体免疫力的增强有着明显的效果。

The immunity of the body can be enhanced through appropriate scientific methods. Drinking tea is a kind of convenient, healthy, and effective way. Tea can help resist bacteria, viruses, and fungi with its health elements. And its Vitamin C can enhance immune function. Some studies also suggested that tea amino acids would build up your resistance. In a word, drinking tea has a noticeable effect on increasing immunity.

消炎杀菌
Anti-inflammatory Sterilization

在中国古代，茶叶便常常被用来消毒伤口。这是因为茶叶中含有的儿茶素类化合物和黄烷醇类能够起到很好的消炎杀菌作用。首先，黄烷醇类相当于激素，能够促进肾上腺体的活动，具有直接的消炎作用。其次，茶叶中的儿茶素类化合物对于多种病原细菌具有明显的抑制作用。茶叶中多酚类化合物和儿茶素类化合物，还可以明显抑制病毒。

Tea was often used to disinfect wounds in ancient China. Tea contains catechins and flavanols compounds for anti-inflammatory sterilization. First of all, flavanol are equivalent to hormones, which can promote adrenal gland activity. Therefore, it has a direct anti-inflammatory effect. Secondly, catechins compounds can obvious inhibit the action of various bacteria. Its polyphenolic and catechin compounds can inhibit the plant virus.

科学合理地饮茶

Drinking Tea Scientifically

 了解茶的功效

Realize the Benefit Effects of Tea

要做到科学合理地饮茶，不仅要把握好饮茶的时间，还要对不同种类和地域的茶叶功效有一定的了解，这样才会有针对性地选择饮用，实现茶叶保健养生效能的最大化发挥。

In order to drink tea scientifically and reasonably, we should not only grasp the time of drinking tea, but also have a certain understanding of the different kinds and regions of tea. so as to drink it in a targeted manney and maximize the health care efficiency of tea efficgcy.

掌握泡茶的技巧
Master Tea Brewing

做到科学合理的饮茶还要掌握一定的泡茶技巧，如果方法不当，也会弄巧成拙，影响茶叶品性的发挥和品饮的实用性。

We also need to master certain tea the brewing skills for scientific and reasonable tea drinking. If the brewing method is not appropriate, it will also be self-defeating, affecting the play of tea character and the practicality of drinking,

茶叶的保存
Preservation of Tea

茶叶是一种比较娇贵的消费品，除了普洱茶可以长年陈放提升价值外，其他茶叶都各有一定的保质期。自然环境中的诸多因素，包括温度、湿度、空气、光线等，都可能引起茶叶成分氧化，造成其颜色发暗，香气散失，味道不佳，甚至发霉，使茶叶保质期缩短，影响到其作为商品的经济价值和饮用口感，甚至还会影响到饮茶者的身体健康。

Tea is a kind of more delicate consumer good. Except the Pu Er Tea can enhance the value after a long time of storing, other teas have a certain shelf life. Many natural factors will may cause the oxidation of tea components, such as temperature, humidity, air, light, etc. The tea color would be darkened, and the tea aroma would disappear. Poor taste, and even mildrew, making tea expire in advance, affecting its economic value as a commodity, drinking taste, and even affecting the health of tea drinkers.

根据茶叶品种、生长条件等不同，茶的保质期也不尽相同。如果在存放过程中采用妥善合理的方法贮存，可以有效保证茶叶质量，减缓茶叶的质变过程，使品茶人能更为安心地体会品饮乐趣。

According to the different vorieties of tea grovoing conditions, etc. the shelf life is not the same. the quality of tea can be effectively guaranteed, the qualitotive change process of tea can be sloned down, and the tea tasting people can feel more at ease.

影响茶叶品质的因素
Main Factors Affecting the Quality of Tea

温度
Temperature

温度能加快茶叶的自动氧化，使得茶叶的香气、汤色、滋味等发生很大的变化。温度愈高，变质愈快，茶叶外观色泽越容易变深变暗。尤其是在南方，一到夏天，气温便会升到40℃以上，即使茶叶已经置于阴凉干燥处保存了，也会很快变质，使得绿茶不绿，红茶不鲜，花茶不香。因此要维持或延长茶叶的保质期，应采用低温冷藏保存，这样可减缓茶叶中各种成分的氧化过程，有效减缓茶叶变褐及陈化。

Temperature can speed up the cuctomatic oxidation of tea. So that the aroma of tea, soup color. And taste have a significantly change. The higher the temperature, the faster the deterioration. The more easily the appearance of tea cdor clarlcer and clarker. In summer, the temperature will increase to above 40, especially in southern China. Even if the tea has been stored in a cool and dry place. Tea will go off quickly in hot weather. Green Tea will not be green. Black Tea will lose its brisk taste. And Scented Tea will not make fragrance. So we need to store tea with low-temperature storage conditions to prolong the quality guarantee period.

湿 度
Humidity

成品茶具有很强的吸湿性，空气中的水分轻易便可被其吸收。如果把干燥的茶叶放在室内，且直接接触空气，只需一天的工夫，其含水量便可到达 7% 左右；放置五六天后，含水量便可上升到 15% 以上。如果在阴雨潮湿的天气里，每暴露置放 1 小时，其含水量就可增加 1%。在温度较高，微生物活动频繁的月份，一旦茶叶含水量超过 10% 时，茶叶便会发霉，色香味俱失，不再适宜饮用。因此茶叶必须在干燥的环境下保存，不能受到水分侵袭。若不慎吸水受潮，轻者失香，重者霉变。一旦发生，不可将受潮茶叶放在阳光下直接暴晒，而应放在干净的铁锅或烘箱中用微火低温烘烤，同时要不停地翻动茶叶，直至茶叶干燥并发出香味方可。

The hygroscopicity of the finished tea is strong. The water in the air can be easily absorbed. If you put the dry leaves in your room to direct contact with the air, their moisture content would reach about 7% within a day. The moisture content will rise to 15% after five or six days. Its moisture will increase by one percent with one more hour exposed.

Tea will get moldy and lose its original taste with above 10% moisture content. So tea leaves should be stored in a dry environment. Damp tea leaves cannot be dried under direct sunshine. It should be baked at a lower temperature and stirred until the tea fragrance comes out.

氧气
Oxygen

茶叶在贮藏过程中，许多化学成分都极易发生氧化作用，导致茶叶陈化和劣变。影响品质的化学成分主要是叶绿素、茶多酚、维生素、类胡萝卜素、氨基酸以及多种香气成分等。茶多酚经氧化后会生成醌类化合物，导致茶的颜色变深，转为褐色，同时此类化合物还能与氨基酸类物质进一步发生反应，使味道发生严重变化。维生素C被氧化成脱氧维生素C后与氨基酸相互作用，生成氨基羟基，这样不仅降低了茶叶营养价值，又使颜色变褐，同时茶的味道也完全丧失鲜美。茶叶中的少量脂肪类物质被氧化、水解后成游离脂肪酸、醛类或酮类，进而出现酸臭味，茶叶香味显陈，而且汤色也会加深。所以茶叶必须隔离氧气，可采用抽真空充氮、密封贮存的方法来保存。

Many chemical constituents will easily suffer oxidation during storage-chlorophyll, tea polyphenols, vitamins, carotenoids, amino acids, various aroma components, etc. Tea polyphenols will produce Quinonoid with oxidation to make the color darker. And this component will also further react with Amino acids to change its original taste seriously. Vitamin C oxidizes, forms deoxy vitamin C, and responds with Amino acids to produce aminocarbonyl group compounds. It will reduce their nutritional value and cause them to lose their freshness and excellent taste. Many fat components will produce free fatty acids, aldehydes, or ketones to make tea leaves sour. So tea leaves should be stored without oxygen. You can keep tea leaves with filling nitrogen in a vacuum package. Some tea leaves should be sealed in storage.

光 线
Sunlight

光线照射也会对茶叶产生不良的影响，光照会加速茶叶中各种化学反应的进行。茶叶中含有少量的类胡萝卜素，这是一类黄色素，成分复杂，是光合作用的辅助成分，具有吸收光能的性质，在强烈光线的作用下极易被氧化。氧化后的类胡萝卜素产生的气味会使茶汤味道改变。茶叶中的叶绿素经光线照射后易褪色，使茶叶的色泽变暗沉。如果把茶叶放在日光下晒一天，则茶叶的色香味都会发生比较显著的变化，从而失去其原有的风味和鲜度。因此，茶叶一定要避光贮藏，日常包装材料也要选用能遮光的为好。

Strong sunlight will also hurt tea leaves. It will speed up chemical reactions. There are a few Carotenoids in tea leaves. It has complex components. It is an auxiliary component of photosynthesis to absorb light energy and is easily oxidized when exposed to bright light. It will produce some odors to make tea with a sour taste. Chlorophyll would decrease to make tea leaves dull. Tea taste will change dramatically to lose its original flavor and freshness in the brilliant sunshine. So tea storage should avoid sunlight.

气　味
Odor

异味污染与茶叶的储存环境有关。因茶叶中含有铁硒化合物和高分子棕榈酸，其性质活泼，很快就能吸收其他物质的气味而改变或掩盖茶叶本来的香气。

如把茶叶和味道浓郁的樟脑丸、香料、药物等放于一处，或把茶叶放在原有气味还未丧失的新木器、新漆器里，用不了几个小时茶叶就会污染上这些物品的气味，轻则让人失去品饮的兴趣，重则完全无法饮用。所以茶叶在贮藏过程中要严防与有异味的物质接触，贮茶容器也必须保持洁净无味。

Smelling pollution will occur in an improper storage environment. Due to active iron, selenium compounds and polymer palmitic acid can quickly absorb other odors to change their original smell. For example, tea leaves will have an awful odor if they store them with camphor pills, spices, drugs, and other odorous compounds. You do not want to drink tea at that time. So avoid contact with other odor substances during storage. And storage utensils should also be clean and odorless.

常用的贮存方法
Most Frequently Storage Methods

有一些常用的贮存方法如真空储存、密封储存、充氮储存、塑料袋储存和冰箱保存等，都可以保持茶叶新鲜。

过期茶叶在品质上已经大大降低，表现在色香味各方面，都有了显著的变化。饮茶前一定要仔细观察，以免喝入变质茶叶影响健康。首先可以看看茶叶外包装袋上的保质日期，如果超过就不要再饮用。过期的茶叶有可能发霉，出现陈放后的霉味，浓烈刺鼻，冲泡后也不再有茶叶的浓醇鲜爽。如果是绿茶的话，过期茶有可能泛红，冲泡后汤色也会变暗，失去原有的黄绿色。

Some common storage methods include vacuum storage, sealed storage, nitrogen-filled storage, plastic bag storage, and refrigerator preservation, which can keep tea leaves fresh.

The quality of expired tea leaves has been greatly reduced. There are notable changes in its color, texture, and shape. It would help if you observed before drinking to avoid expired drinking tea. First, you can check its guaranteed date on the packaging bag. Do not buy those bulk tea leaves without printing the expiration date. Out-of-date tea leaves will be brittle and dry rather than lush and aromatic. It probably had mold on it. It will lose its mellow and brisk taste. Those expired Green Tea leaves will be even with a red edge. Its brewed soup will be darker. Its original yellowish-green color will also disappear.

茶风篇
Spread in the World

茶风茶俗承载了历史，成为很多国家民族文化的一部分，饮茶也是他们生活的一部分。体会异域的茶情，也就感悟了他国的文化；玩味异域的茶趣，也就了解了茶中别样的真谛。

The tea culture and customs carry history and have become a part of the ethnic culture of many countries. Drinking tea is also a part of people's lives in these countries. By experiencing the exotic tea culture, one can also appreciate the culture of other countries; by savoring the excellent taste of tea, one can also understand the unique essence of tea.

茶风篇

Spread in the World

随着时间的推移，茶逐渐流传开来，并成为中国文化不可或缺的一部分。从古至今，茶在中国文化中扮演着重要的角色。茶作为中国文化的一部分，是一种拥有多年历史和丰富文化遗产的饮品。

With the passage of time, tea gradually spread and became an indispensable part of Chinese culture. Tea has played a role in Chinese culture from ancient times to the present. As a part of Chinese culture, tea is a beverage with a long history and rich cultural heritage.

在《诗经》中，就有关于"茶"的记载，诗中描述了采茶、制茶、饮茶的情景。此外，在《史记》《神农本草经》等古籍中也有关于茶的记载，据《神农食经》记载："坚持长时间饮茶，能让人精神饱满、心情愉悦。"说明在那个时期，茶已经成为人们生活中的重要组成部分。

The Book of Songs has a record of "tea," which describes the scene of picking, making, and drinking tea. In addition, there are also records about tea in the Historical Records, the Shennong Ben Cao Jing, and other ancient books. According to the Shennong Food Classic, "Persisting on drinking tea for a long time can make people full of spirit and happy." It shows that tea has become an essential part of people's life at that time.

依照《广雅》中所记述的，在荆州、巴州地区，人们采摘茶叶来作成茶饼，对于那些芽叶较老的茶，做茶饼时要添加米糊搅拌才行。假如要煮茶饮用，应先将茶饼烤成红色，然后捣碎置于陶瓷器皿中，再加入沸水冲泡，或者是把葱、姜、橘子放在一起搅拌浸泡。饮用这样做的茶可以醒酒，让人精神振奋不思睡眠。

According to Guang Ya, in Jingzhou and Bazhou areas, tea leaves are picked to make tea cakes. Those with aged leaves need to be mixed together when making tea cakes. If you want to make tea for drinking, you should first roast the tea cakes into red, then mash them into ceramic containers, and then add boiling water to make them. Or you can put onions, ginger, and oranges together to stir and soak them. Drinking the tea made in this way can sober up people's spirits and make them not want to sleep.

三国两晋南北朝时期，中国处于分裂状态，政治动荡，但茶文化依旧在这个时期得到了蓬勃发展。两晋南北朝时期，文人雅士们对茶的品鉴更加注重，出现了不少有关茶的诗词歌赋。

During the Three Kingdoms, the Jin Dynasty, and the Northern and Southern Dynasties, China was divided and in political turmoil, but Tea culture still developed in this period. During the two Jin and Northern and Southern dynasties, scholars and scholars paid more attention to tea tasting, and many poems and songs about tea appeared.

西晋左思的《娇女诗》大意说："我家有两个乖巧的女儿，长得非常白皙，小名叫纨素，口齿极为伶俐。姐姐叫蕙芳，眉目美如画，在园林里蹦蹦跳跳，果子未成熟就摘下来。她们爱花哪里顾得上风

和雨，跑着进进出出百余次，看到煮茶心里就兴奋，对着茶炉往里面吹气。"

In Zuo Si's Poems of a Beautiful Girl in the Western Jin Dynasty, it is said that I have a cute daughter in my family, who is very fair and nicknamed Wan Su. She Loves flowers and doesn't care about the wind and rain, running in and out more than a hundred times that day. I was excited when I saw the tea brewing and then blew air into the stove.

壶居士在《食忌》中表示：经常饮茶，身轻体健，就如同飘飘欲仙；茶和韭菜一起食用，可以增加身体的重量。

Hu Jushi said in his book Food Taboo: Drinking tea regularly makes one feel light and healthy. Eating tea with leeks can increase one's body weight.

《茶陵图经》里也有记载：茶陵，即陵谷中生长着茶树的意思。

According to the "Chaling Classic," "Chaling" means tea trees grow in hills and valleys.

茶经八之出

Chapter 8: Spread in the World

茶风篇
Spread in the World

茶,源于中国,传向世界。从古代丝绸之路、茶马古道、茶船古道,到今天丝绸之路经济带、21世纪海上丝绸之路,茶穿越了历史、跨越了国界,深受世界各国人民喜爱。

Tea originated in China and spread to the world. From the ancient Silk Road, the Tea Horse Ancient Road, the Ancient Tea Boat Road, to today's Silk Road Economic Belt, and the 21st century Maritime Silk Road, tea crosses history and borders, and is deeply loved by people all over the world.

茶叶是中国老百姓再熟悉不过的饮品。自古以来,中国茶叶随着丝绸之路传到欧洲,逐渐风靡世界,与丝绸、瓷器等一同被认为是共结和平、友谊、合作的纽带。

Tea is the most familiar thing to the Chinese people. Since ancient times, Chinese tea has spread to Europe along the Silk Road and gradually became famous worldwide. Tea, silk, and porcelain are the bonds of peace, friendship, and cooperation.

在世界上传播
Spread in the World

中国茶叶向海外传播的历史已久，最早可追溯到南北朝时期。那时的中国商人便开始在与蒙古的边境线上，通过以茶易物的方式向蒙古输出茶叶。大约到了汉代，中国茶叶开始传入日本。到隋唐时期，在对外开放政策的支持下，以及丝绸之路开通的基础上，边境贸易得到发展壮大，中国商人以茶马交易的方式，使茶叶经回纥及西域等地向外输送，中途辗转西伯利亚，送往西亚、北亚和阿拉伯国家，最终抵达俄罗斯和欧洲各国。

The history of spreading Chinese tea overseas is very long. It may be traced back to a period as early as the Northern and Southern Dynasties. At that time, Chinese traders had begun to export tea to Mongolia by barter trade. During the Han Dynasty, Chinese tea started to be shipped to Japan. With the Chinese open policy, tea leaves were exported along the Silk Road in the Sui and Tang Dynasties. Border trade was developing rapidly at that time. Chinese merchants transported tea through Hui and Xicheng land by means of tea hase trading Tea had been shipped to Siberia, west Asia, north Asia, Arab countries, Russia and Europe.

明代郑和下西洋，把茶叶经过海路传播到东南亚和波斯湾一带；自17世纪起，茶叶相继被传到荷兰、英国、法国、德国、瑞典、丹麦、西班牙等欧洲国家；18世纪，饮茶风俗已经传遍了整个欧洲，欧洲殖民者又将中国茶带到美洲大陆及大洋洲的英、法殖民地；到19世纪，中国茶就已经几乎遍及全世界，成为众所周知的饮品。

During his voyage, Zheng He spread tea leaves to Southeast Asia and the Persian Gulf. Tea has been shipped to European countries such as Holland, Britain, France, Germany, Sweden, Denmark, and Spain since the 17th century. The tea-drinking custom spread through Europe in the 18th century. European settlers also brought tea to the American continent and Oceania. So Chinese tea has been distributed worldwide and has become famous for drinking since the nineteenth century.

茶马古道
Ancient Tea Route

起源与发展
Origin and Development

　　位于中国西南地区的茶马古道是以马队为交通工具，进行民间商品贸易的通道。古道起源于古代西南边疆的茶马互市，最初的线路即青藏线始于公元7世纪。那时居住在青藏高原的吐蕃民族崛起，南下到中甸境内的金沙江上了建造了一座铁桥，从此打通了云南向西藏输送茶叶的往来之路。青藏线在唐朝时期十分繁荣，行走在古道上的马帮为茶叶的传播做出了不朽的贡献。

　　People rode horses on the ancient tea route in China's southwestern region for commodities. The ancient tea road originated from the tea-horse trade center at China's southwest border. The initial Qinghai-Tibet line dates from the 7th century. Tubo people made significant development and went down to the Jinsha River. They built an iron bridge to open up the tea shipping route. The Qinghai-Tibet line was prosperous during the Tang Dynasty. Caravans made a monumental contribution to the spreading of tea.

茶马互市
Tea-Horse Trade Center

在中国的西部地区，长期以来生活着很多马背上的民族。对于这些民族的人们来说，每天吃肉喝奶，缺乏必要的维生素的摄入，所以饮茶就变得格外重要。所以有一句俗话就叫做："一日无茶则滞，三日无茶则病。"而对于广大的汉族地区来说，盛产茶叶，但是却没有马匹。在过去，战场上骑兵是主导力量，汉族地区的统治者为了增强自己的军事实力，就采取控制对少数民族供应茶叶数量的办法，以抬高茶叶价格，用少量的茶叶来换取更多的马匹。这样茶叶就在控制周边地区中起到了重要作用，这就是"以茶治边"。清朝时，中央政府对于茶叶的流通控制不力，茶叶的价格越来越便宜，不得已之下，雍正十三年（1735）的时候，茶马法被正式废弃。

Many nationalities on horseback eat meat and drink milk daily in western China. Most of them have a deficiency of essential vitamins. So drinking tea has become particularly important and necessary. It probably gave rise to the proverbial saying that you would be sluggish without tea one day and be sick without three days. Han nationality had abundant teas but without horses. In the pasb, cavalry was the dominant force on the battlefield, and the nders of the Han region took antrol over ethnic minorities in order to enhance their milltaly strength the way to supping the quantity of tea is to pick up the price of tea. Exchange a smell amount of tea for more horses. They traded with each other. Tea played a vital role in taking control of areas surrounding Tibet area. The policy of harnessing the borders with tea was implemented. Due to the tea trade's under-control from the central government in the Qing Dynasty, tea prices had become cheaper and cheaper. Tea-Horse Law was abandoned in thirteen years of Yong Zheng.

主要线路
Main Line

茶马古道是一个广义的统称，其主要线路共有三条：包括青藏线、滇藏线和川藏线。其中的青藏线始于唐朝时期，历史最为悠久；而川藏线则对后代的影响最大，也最为著名。

All tea business lines traded with tea are called the Ancient Tea Route. There are three main lines: Qinghai-Tibet, Yunnan-Tibet, and Sichuan-Tibet. The Qinghai-Tibet line began in the Tang Dynasty and has the most extended history. The Sichuan-Tibet line is the most famous and has the most profound influence.

到达终点
Reach the End

茶马古道从走向看来可以称其为一条路，其实际上却是一个庞大的道路群，局部线路细密如织，犹如江河的主脉与细密支流最终汇入大海。其主脉主要有两条，即川藏古道与滇藏古道。最终这两条线路都在西藏东部的昌都会合，然后经洛隆、嘉黎等县到达终点——拉萨。

The ancient Tea Route also can be called a road. It is a vast road group. This tea route provides a comprehensive road network for spreading tea. There are two main branches: the ancient Sichuan Tibet Road and the Ancient Tea And Horse Road in Yunnan-Tibet. They would meet in the city of Changdu. Then pass the town of Yulong and Jiali to Lasa.

其实，拉萨既可以说是终点又可以说是起点，这是因为从更广泛的意义上来看，茶马古道继续向西延伸，从拉萨出发，经江孜、日喀则，朝西南方去向到中亚、西亚和南亚。事实上，自古以来，很少有人能够真正走完这条长达万里的艰险古道。

However, Lasa is the finishing place of the Ancient Tea Route in China and the beginning of the tea route to the west. It is long from Lasa to Central Asia, Western Asia, and South Asia. In tact since ancient times Few people can walk or travel far on this difficult ancient road for thousonds of miles.

传入日本
Introduced into Japan

与佛教同行
Introduced with Buddhism

唐朝时期，日本僧人最澄禅师来到中国浙江天台山的国清寺研究佛学，其间接触到茶且十分喜爱。回国时，他带回茶籽并传播到日本的中部和南部地区。南宋开庆年间（1253-1259），又有数名日本佛教高僧来到浙江径山寺研习佛学，将径山寺的茶宴活动和"抹茶"的制法传播到日本，对日本茶道的兴起产生了启发和巨大的促进作用。

During the Tang Dynasly. The Japanese monk Saicho went to Guoqing Temple in the Tiantai Mountains in the Zhejiang Province to study Chinese Buddhism. He came into contact with tea and was fond of it. When he returned to Japan, he brought some tea seeds and spread them in Japan's central and southern parts. Other Japanese monks brought back some tea utensils in the Kaiqing peried of the Southern Song Dynasties (1253-1259) and spread Japan's tea banquet and powder tea processing method. It contributes a lot to the sustainable development of the Japanese Tea Culture.

在日本，茶因佛教的发展而被引入并传播，佛教教义对于日本茶道文化精髓的形成也产生了重要影响。总而言之，日本的茶道呈现出"佛中有茶，茶中有佛，佛离不了茶，茶因佛而兴"的文化特色，故此有"花佛一味"或"禅茶一味"之说。

Tea gradually became famous as the Buddhism culture was introduced into Japan. Buddhist doctrine also plays a vital role in establishing Japanese tea culture. In conclusion, tea is in Buddhism, and Buddhism is in tea. They can't be separated from each other. So there is a combination of tea and Zen.

茶籽的传播
The Spread of Tea Seeds

中国茶种最早传播到的地方是东亚邻邦——日本，据《日吉神道密记》记载，公元805年，从中国学佛归来的最澄禅师将茶籽带回日本，并栽种于日本贺滋县的日吉神社的旁边，从而使那里成为日本最古老的茶园。至今在京都的比睿山东麓还立有一块"日吉茶园之碑"，其周围至今仍有茶树在茁壮生长。日本茶业经过引种、扩种、再植等长久而缓慢的发展历程，至19世纪以后，逐渐步入了上升时期。

Those tea seeds were first spread to the neighboring country Japan in East Asia before. According to historical records of The secret of Jilin Shinto, Japanese monk Saicho brought some tea seeds back to Japan and spread them in the central and southern parts of Japan. They planted tea beside the shrine, making it the oldest Japanese tea garden. There are still some tea trees around it. Japanese tea culture and industry experienced a slow development course until the 19th century. Then gradually step into the rising stage.

日本茶道
Japanese Tea Ceremony

日本的饮茶风尚是由贵族社会逐渐向广大群众普及的。日本的茶道源于中国，同时也具有日本民族特色，它有自己的形成、发展过程以及自身特有的内蕴。

Japanese tea drinking had been given to the popularization gradually from aristocratic society. It originated in China and with Japanese national characteristics. Therefore, the Japanese tea ceremony has its own pormation development process and a unique implication.

16世纪末，日本茶道的集大成者千利休汲取并继承了历代茶道精神，正式创立了日本茶道。日本茶道以"和""敬""清""寂"四字为宗旨，内容精炼而内涵丰富。它以日常生活行为为基础，与宗教、伦理、哲学和美学艺术等内容融合在一起，形成了一项综合性的文化艺术活动。饮茶不仅仅是一项饮食活动，更要通过茶事活动来学习礼仪、陶冶情操、培养审美观和道德观。

The master Sen Rikyu carried on the tea ceremony spirit before and formally created the Japanese Tea Ceremony. Its aims are peaceful, respectful, pure, and quiet. Its content is refining and colorful. The Japanese tea ceremony integrates religion, ethics, philosophy, and aesthetics. Drinking tea is a diet activity for learning etiquette, edifying sentiment, and developing aesthetics and morality.

来到欧洲
Spread to Europe

海上之路
On the Sea Route

唐宋时期，中央政权积极推行对外开放的国策，使得茶叶得以流传到世界更广泛的地区，也为"海上茶叶之路"的形成与发展奠定了坚实的基础。

The opening-up policy was actively promoted in the Tang and Song Dynasties. As a result, tea could be spread to the broader region of the world. It also laid a solid foundation for forming and developing the Tea Road on the sea.

茶叶的海上之路最早是通往中国的海上近邻——日本和朝鲜，主要线路是从江苏、安徽、浙江和福建等茶区出发，由宁波、扬州和泉州的港口入海。

The Marine Tea Road led to our neighbor countries, Japan and Korea, at the earliest. The main routes started in tea-producing areas such as Jiangsu, Anhui, Zhejiang, and Fujian provinces. from the port of Ningbo. Yang zhou and Quanzhou into the sea.

此外,茶叶输往海外其他国家的主要海路有两条:一是从江西、浙江、福建茶区出发,经宁波、泉州和广州的港口入海,直接横跨太平洋运往美洲;第二条是从中国的茶区输往南洋,再驶过印度洋、波斯湾和地中海等地后销往欧洲各国。

Besides, there are two more main sea routes for exporting tea to other countries. One is from Jiangxi, Zhejiang, and Fujian to America. The other is from the Chinese tea-producing area and sweeps across the Indian Ocean, the Persian Gulf and Mediterranean sell to European countries.

功不可没的传教士
A Crucial Contribution from Missionary

公元851年,阿拉伯人苏莱曼在《中国印度见闻录》一书中介绍了中国广州的情况,其中就特别提到了茶叶。公元14~17世纪,中国的茶叶经由陆路输往中亚、波斯、印度西北部和阿拉伯等地区,再通过阿拉伯人首次被传到西欧。这一时期,欧洲的传教士也开始来到中国传教,他们不但为中西方文化的沟通交流搭建起了桥梁,

也将中国丰富的物产，特别是茶叶介绍到了欧洲。

An Arab people, Soliman introduced the Chinese city of Guangzhou in his book CHINA AND INDIA SKETCHBOOK. He specifically mentioned tea in the book. Chinese tea was exported to Central Asia, Persia, Northwestern India, Arabia, and Western Europe by land from the 14th century to the 17th century. During this period, the European missionaries also went to do missionary work in China. They not only built a communication bridge between China and the West and brought rich Chinese natural resources, such as tea, back to Europe.

荷兰人的优势
The Advantages of Dutch

在17世纪初的时候，欧洲的荷兰是世界上的商业大国，也拥有着世界上无与伦比的海上力量。荷兰的商船频繁地行驶在辽阔的大海上，进行着贸易往来。聪明的荷兰人从澳门装运中国绿茶，然后运输到欧洲。由于只有荷兰人掌握着茶的资源，因此茶在当时的欧洲非常名贵。茶和贵族紧紧地联系在一起，也和奢侈风尚紧紧地联系在一起。以至于在当时的欧洲，茶仅仅在宫廷贵族和豪门世家之间出入，饮茶成为一种身份的标志和骄傲。

The Dutch was the biggest commerical country around the worlol, and to trade with other countrnes had a world-class trading power in the early 17th century. The Dutch merchants frequently ran in the vast sea. Those smart Dutchmen shipped Chinese Green Tea from Macao to Europe. Because only Dutchmen could grasp tea resources, tea was famous and expensive in Europe then. Tea and nobility are closely linked with each other. It was a kind of prevailing extravagance. In Eurppe at that time tea only existed in aristocratic societies. Tea drinking has become a status symbol and pride.

万里茶路
Tea Route across a Great Distance

中俄两国于清雍正五年（1727）签订互市条约，开始以中俄边境重镇恰克图为中心进行通商贸易，茶叶便是其中重要的商品。商人们将茶叶用马匹运到天津，然后再用骆驼运送，驼队穿越茫茫大草原和万里大沙漠，最终抵达中俄边境口岸恰克图进行交易。俄国商人们将茶叶贩卖到西伯利亚伊尔库兹克、乌拉尔、秋明等地区，甚至一直运送到遥远的莫斯科与圣彼得堡。

The frontier trade treaty was signed between China and Russia in the fifth year of Emperor Yongzheng (1727). The starting point was the border town of Kyakhta. Tea was one of the essential commodities. Traders came to barter with horses to the city of Tianjin. Then transfer goods with camels. Next, they would cross the vast prairie and great desert. Finally, where they eventually reached Kyakhta. Russian traders sold tea to Siberia, Ural, Tyumen, etc. Tea had even been sold to Moscow and St Petersburg.

这条贯穿南北、水陆交替的运输之路从福建崇安（今武夷山市）出发，途经江西、湖北、河南、山西、直隶（今河北省）及内蒙古，最终到达乌里雅苏台（今蒙古人民共和国）的恰克图，全程约4600公里（9200余里），被人们称为"万里茶路"。这条茶路持续兴盛了150余年，是一条可与"丝绸之路"相媲美的辉煌繁荣之路。

This transport route runs through the north-south alternating waterway road. It was from Wuyi Mountain City to Kyakhta by Jiangxi, Hubei, Henan, Shanxi, Hebei, and Inner Mongolia. The whole route is 4600 kilometers, which was always called Wanli Tea Route. It flourished for more than one hundred and fifty years. It is a brilliant and prosperous road that can comparable with the Silk Road.

茶在英国
Tea in Britain

嫁入英国的中国茶
Chinese Tea Married into Britain

1662年，英国国王查尔斯二世与葡萄牙公主凯瑟琳结婚。凯瑟琳公主婚前就是饮茶爱好者。因此，她出嫁后便把茶叶带到了英国。她在宫廷中用茶招待王室贵族，逐渐使茶的名气流传开来，带动了全国饮茶的风气。

后来，英国将茶传播到殖民地及德国、法国、瑞士、丹麦、西班牙、匈牙利等欧洲国家。

In 1662, king Charles II of England married the Portuguese princess Catherine of Braganza a tea drinker, who brought a small tea chest as part of her dowry. Then she entertained guests with tea. The fame of tea gradually spread out. Then the whole country started to drink tea. After that, Britain began to ship tea to its colony and Germany, France, Switzerland, Denmark, Spain, Hungary, and other European countries.

神奇的"药水"
Magic Medicinal Liquid

茶最初进入英国，曾被当做一种包治百病、神奇昂贵的饮料，并且具有浓厚的异国情调。对于初识茶叶的大多数英国人来说，茶树更像是神话传说中的一株仙草。

Tea was once treated like an expensive magic drink. People claimed it was a miracle cure with a solid foreign tone. For most British people who are new to tea, the tea tree was a kind of magic herb in myths and legends.

英国有一位经营咖啡馆的先生名叫加乐维(S.Garaway)。1658年，他开始在自己的店里售茶。他在张贴的广告中列举出茶叶的十几种药用功效，其中包括可以医治脑卒中、肾结石、尿道结石、脱水、水肿、坏血病、腹泻或便秘、头痛、嗜睡、多梦、记忆力减退等。此外，他还列举了茶还可以增进食欲、补充营养、消积去淤、利肾清尿。饮茶的方式也多种多样，可以用开水冲饮，还可以加入牛奶、糖或者蜂蜜。

S.Garaway, a coffee shop in Britain, started to sell tea. In 1658 He listed a dozen medicinal efficacy of tea, including stroke, stones, urinary sand, dehydration, edema, scurvy, abdominal discharge or constipation, headache, drowsiness, dreaminess, memory loss, etc. Furthermore, tea can enhance your appetite, supply nutrition, remove food retention, and improve kidney function. Drinking tea methods are also varied. To taste, you can brew with boiled water or add milk, sugar, and honey.

全民走私
Smuggling Tea

茶叶作为奢侈品引入英国后，政府自然要课以重税，最高时竟达货值的20%。尽管税收如此之高，东印度公司还是因茶叶贸易而获得了巨额收益。

Tea was introduced into Britain as a luxury good. The government imposed a heavy tax on it. The tax even reached twenty percent of the total value. Although with the high tax rates, The East India Company still made an immense fortune with tea trading.

东印度公司刚刚成立的时候，主要是进行香料与胡椒等物的贸易，随着英国的茶叶消费市场越来越大，其主要贸易商品渐渐变成了茶叶。东印度公司对茶叶经营实行了垄断，它所提供的中国茶叶不但价格奇高，在数量上也远远不能满足英国市场的需要。至18世纪中期，英国民众消费的茶叶有将近一半源于走私，来自于荷兰倾销的茶叶，价格在当时十分便宜。于是，随着饮茶风俗逐渐普及，除了美洲的烟草与法国的白兰地，整个英国海岸线都在忙碌着走私中国的茶叶。

Its main business was spices and pepper trading of the beginning of the East India Company. With the rapid development of portable tea, tea became its primary consumer good. The East India Company had a monopoly on tea. Chinese tea was expensive and could not meet the UK's requirements. Nearly half of the tea was smuggled from the Netherlands since the mid-18th century. The price was meager. With the gradual popularization of tea-drinking customs, Britain was busy smuggling Chinese tea, American tobacco, and French brandy.

运茶比赛
Shipping Contest

一年一度的中英运茶大赛促进了帆船时代海上航运的发展，同时也是中英茶叶交易的历史见证。当时的"卡蒂萨克"号是大海上跑得最快的运茶船。在1872年从上海开往伦敦的运茶大赛上，"卡蒂萨克"号与"塞莫皮莱"号展开了激烈的竞争。

The annual tea shipping contest promoted the development of the shipping industry in the Big Maritime Navigation Era. It was a historical testimony of the tea trade between China and Britain. The Cutty Sark was one of the fastest sailing ships ever built. In 1872, there was A fierce competition contest was held between the Cutty Sark and Thermopylae.

这场运茶比赛持续了4个月之久，是此类比赛中最著名的一次，同时也是最后一次。此后，运茶的帆船逐渐退出了历史舞台，因为"蒸汽机"的出现开启了航海的新纪元。

The match lasted four months. It was the most famous and the last game of such contests. After that, the sailing ship gradually withdrew from the historical stage. The appearance of the steam engine opened a new modern era of navigation.

另辟产地
Creating a New Producing Area

亚欧的茶叶贸易起源于中国。由于当年航运周期比较长，以及欧洲人的口味偏好，英国人选择了全发酵的红茶。红茶的起源应该是武夷山的正山小种。

The tea trade of Asia-Europe originated in China. Britishers were fond of the fully-fermented Black Tea due to the longer shipping cycle and taste. The origin of Black Tea was Wuyi Mountain Lapsang Souchong.

早期东印度公司垄断茶叶贸易，并由于茶叶贸易带来了暴利，英国人一直想引进茶叶种植，但是因为地理条件的限制而没能成功。

The East India Company held a monopoly on the tea trade. It could yield British men valuable profits. So the British tried hard to introduce teas. But they had failed due to the limitation of geographic conditions.

后来经过百般周折，终于在其殖民地印度成功地种植了茶树。北印度的海拔和气候十分适宜茶树生长，因此后来印度与斯里兰卡成为重要的红茶出口国。一直到今天，印度仍然是亚欧茶叶贸易第一大国。

After a lot of twists and turns set backs Eventually, they planted tea trees in the British colony of India. The elevation and climate of Northern India are pretty suitable for Black Tea's growth. Later, India and Sri Lanka have become significant exporters of Black Tea. As a result, India is the largest tea trader in Asia and Europe today.

异域茶情
Tea Culture of Foreign Lands

异域风情的茶文化
Exotic Tea Culture

  茶起源于中国，但是随着自古以来的文化交流，已经传播到世界各地。邻邦日本更是以茶道闻名天下；朝鲜、韩国也有自己沿袭了几个世纪的茶礼；英国人则有雷打不动的下午茶风俗；摩洛哥、土耳其都有别具风味的茶品和独特的饮茶习惯。茶风茶俗承载了历史，成为很多国家民族文化的一部分，饮茶也是他们生活的一部分。体会异域的茶情，也就感悟了他国的文化；玩味异域的茶趣，也就了解了茶中别样的真谛。

  Tea originated in China. As cultural communication expands, it has spread to the world. Japan is famous worldwide for its tea ceremony. North and South Korea inherited tea etiquette for centuries. Britain maintained its unshakeable tea custom. There are some unique tea products and drinking customs. Tea customs contain a lot of important history and have become a part of national culture. Drinking tea is also a part of their life. To understand life's truth, you can experience other countries with different tea cultures.

朝鲜、韩国茶礼
Tea Etiquette of North Korea and South Korea

清、敬、和、乐
Relax Humble, Gentle Cheerful

  朝鲜茶礼讲究清、敬、和、乐四字真谛。清,指不论是泡茶人或是喝茶人都应该带着一种轻松、愉快的心情。敬,指喝茶时应怀着一颗谦卑的心去对待。和,即平和、和气,指泡茶人或喝茶人在享受茶道时应该摒弃所有的杂念和烦恼,用平和的心去细细品味。乐,即快乐、欢乐,这是茶给人们带来的礼物,希望所有的人喝过精心冲泡过的茶后都能忘掉烦恼。

  The true meaning of the Korean tea ceremony pay attention to relaxed, humble, gentle, and cheerful. No matter the brewer or guest, they should be relaxed to enjoy tea. They should stop all distracting thoughts and annoyances to experience a peaceful mind. Tea can also bring a lot of cheer to people.

高丽五行茶礼
The Five Elements Tea Ceremony of Korean

　　五行茶礼是韩国的传统礼仪，也是国家级的最高茶礼。其参加人数众多，规模宏大，有着丰富的内涵，富于诗情画意和民族风情，是为了纪念神农氏而举行的祭祀仪式。

　　The Five Elements Tea Ceremony is traditional etiquette and national etiquette. Most people attend a tea ceremony on a massive scale with a long history and rich content. The tea ceremony is full of poetic and rich ethnic customs. It is a sacrificial ceremony in memory of Shennong.

日本茶道
Japanese Tea Ceremony

和、敬、清、寂
Peaceful, Respect, Pure and Quiet

日本茶道的精神可以归结为四个字：和、敬、清、寂。由茶道高僧千利休确立。与朝鲜茶道相类似，均有"和、敬、清"三字。由此看来两国不仅为邻邦，而且在对茶的理解和造诣上也十分相似。

The Japanese tea ceremony spirit can be summed up in four words peaceful, respectful, pure, and quiet. It was established by the eminent monk Sen no Rikyu. It is pretty similar to the Korean tea ceremony. The two countries are not only neighbors, but also very similar in their understanding and accomplishments of tea.

和、敬，即和气、尊敬。狭义上理解，便是煮茶人与喝茶人之间和睦相处，互相尊敬。从广义上讲，则是希望社会、国家，乃至世界安定和平。清、寂，即清静、幽寂，表示煮茶、喝茶的环境，陈设清静幽寂，使人能够凝神沉思，从而达到茶道所提倡的境界。

Peacefulness and respect mean that people should drink in peace and harmony. So may the world be full of peace and love in broad terms. Pure and quiet can be described as the environment of brewing and drinking tea and the furnishings are pure and quiet. So that people can concentrate on meditation and reach the realm advocated by the tea ceremony.

日本茶道的发展
Development of the Japanese Tea Ceremony

公元9世纪，茶文化随佛教从中国传入日本并发展壮大。日本制茶方式源自唐朝，发展至今已有一千多年的历史。而日本茶道则是15世纪中后期开始出现，到了1586年日本丰臣秀吉执政时代，千利休被授予茶道高僧的美誉，始创茶道的精神原则并集茶道之大成。明治维新之后，茶道有所衰落，直到第二次世界大战之后，日本经济有了明显的复苏和发展，茶道才又开始兴盛和流传。

Tea culture was introduced in Tapan in the 9th century with Buddhism from China. The Japanese way of making tea originated from the Tang Dynasty, with a thousand years of history. The Japanese tea ceremony has existed since the later 15th century. In 1586, during the reign of Toyotom Hideyoshi, sen no Rikyu was appointed aso high priest of the tea ceremony and he initiated the spiritunl principles of the tea aremony and integrated its achievemens. The tea ceremony declined after the 1868 Meiji Restoration. Then it recovered and developed after the Second World War. The tea ceremony began to flourish and spread.

由于烹茶工艺本身十分复杂，日本国内各个流派各自发展茶道，生机勃勃。其中最为盛行的是抹茶道和煎茶道。抹茶道是将采摘的绿茶经过蒸汽杀青后晒干碾碎再磨成粉，这是延续了中国唐宋时期的煮茶方法。泡制时用竹茶帚搅拌，使茶沉入水中，再用竹舀将茶水舀出饮用。煎茶道是将茶叶经过煎炒加工后饮用，源于中国明清时期的饮茶法。

The craft of brewing is very comples, various schools in Japan have developed and vibrant. There are different branches in various genres. Among the most popular ceremonies are the matcha tea ceremony and the leaf tea ceremony. People pluck and steam Green Tea leaves, then crush them into powder for the powder tea ceremony. It is a continuation of the brewing method of the Tang and Song Dynasties. When processing with bomboo tea broom stir, make the tea sink into the water, and then use bamboo scoop to scoop out the tea and drink. The pan-fned tea ceremony refers to drinking tea leaves after frying and processing, Whichderives from the drinking method in China during the Ming and Qing Dynasties.

一期一会
Tea Meeting

一期一会，从字面上理解就是定期展开一些关于茶的系列活动。一期一会包括水、饭、谈、茶四大步骤。其活动的意义在于使参与者的精神与心灵受到一份洗礼，从活动中思考人生的悲欢离合，思考其自身的意义，以及自身肩负的重担，变压力为动力。活动鼓励人们领悟珍惜，即珍惜身边的亲朋好友，珍惜眼前人，珍惜自己现在拥有的一切，珍惜"每日"的缘，以及与"每个人"的缘。

The tea meeting includes four significant items water, meal, chat, and tea. The most crucial profound significance is that the spirit and will of participants can be baptized. People can think about your vicissitudes of life. It can encourage people to cherish those who love them and the person they love.

茶禅一味
The Combination of Tea and Zen

在日本，茶道与禅道关系十分密切。因将茶在日本广泛种植而被称为"日本茶祖"的荣西禅师在中国学禅时，每逢疲惫便用茶水调理身心。回国后，他不仅将中国佛法普传而且把吃茶的风气带给了日本人。荣西禅师曾经说："茶，乃调整心律，强化内脏，平静心灵之良药也。"茶禅一味，追求的是一种喝茶时的意境，喝茶时对自己心灵的一种磨砺。按照传统说法，若是一个人没有"佛缘"，喝起茶来也不能达到二者一体的认知。茶禅一味是茶人基于佛法的一种领悟，是对于茶本身的思考，是心灵和行动的默契，真正做到了表里如一。茶禅一味，让人能保持用平常心态面对所有的烦恼和困难，做到凡事都会去认真思考，注意生活中的点点滴滴，真正领悟"忍"的含义，等等。

Tea has a very close relationship with Zen in Japan. Japanese Tea Saint Eisai Zen-Master refreshed himself with tea when he was tired. He brought back Chinese Buddha dharma and tea customs to Japan.

The combination of tea and Zen pursuits the ideal state. According to tradition, the drinkers cannot obtain sublimation without fate to Buddhism. It is a kind of experience and thought of tea. Drinkers should accomplish this consistently. With the mind at ease, people relax and are happy.

日式茶具
Japanese Tea Utensils

由于日本的茶道源于中国，所以日本茶道使用的茶具和中国功夫茶使用的茶具相似。茶具总共分为四大件，包括煮水用的风炉，有盖的大钵，泡茶用的茶壶和盛水用的茶碗。茶具的质地也分很多种，如陶瓷、漆器、铜器、木器、竹器等。除了四大件之外，还有许多小件各有各的用途，如研磨茶叶的茶磨，夹炭用的火箸，放冷水用的水注，清洁茶具用的水翻，取茶用的茶勺，盛茶叶用的茶罐等。

Due to the Japanese tea ceremony in China, those tea utensils are similar to Chinese tea wares. It is divided into four parts. They are a blast furnace, a big cover bowl, a tea kettle, and a tea bowl. There are also many materials, such as ceramics, lacquerware, bronze, wood, bamboo, etc. In addition to the four parts, some small tools have their use, such as tea mills, tongs, water vessels, teaspoons, tea canisters, etc.

茶风篇
Spread in the World

土耳其茶事
Turkish Tea Story

土耳其红茶
Turkish Black Tea

土耳其人爱喝茶，尤其是红茶。作为一个深受西方文化影响的国家，土耳其人喝茶所用的茶具区别于中国功夫茶，使用形状类似窄腰阔肚花瓶的玻璃杯、小匙、小铜碟。承载玻璃杯的小碟旁一般还要放3粒方糖。

Turkish people love drinking tea, especially Black Tea. Turkey is strongly influenced by western culture. Their tea utensils are different from Chinese Gongfu Black Tea wares. Its shapes are similar to the narrow waist glass cup. You must put a small spoon and three sugar cubes on a copper saucer.

煮茶的方法也有很大差别。一般使用大小两个铜质的茶壶，装

有水的大茶壶放置在木炭火炉上加热，小茶壶内投入适量茶叶放置于大茶壶上。待大茶壶内水煮沸后，冲入小茶壶内。约5分钟后，再将小茶壶中的较浓的茶水依照个人口味酌量倒入玻璃杯中，再注入沸水，并依个人习惯加入方糖即可。饮用时用小匙轻轻搅拌均匀，茶汤口感甘醇，香气怡人。

There is also a big difference between brewing methods. Turkish always use two cooper tea bowls. Then heat the water kettle on the stove. Put some tea leaves into the small kettle. Then pour boiled water into the kettle from the bigger one. Then add sugars according to your taste. Use the small spoon to stir gently. The taste of tea is mellow and sweet, with a pleasant fragrance.

特色茶室
Characteristic Teahouse

各式各样的茶室遍布土耳其的大街小巷，甚至连人迹稀少的角落也不例外。每当华灯初上，大大小小的茶室里总是坐满了人，与一般咖啡店不同的是，大家总是喜欢围成圈坐，喝喝茶，尝尝点心，聊聊天，有说有笑。忙碌了一整天的紧张心情在此时此地得到放松和舒缓。

Various tea houses can be found all over Turkey. They are full of guests every night. People sit in a circle and taste local snacks. As a result, they can feel relaxed and peaceful.

英式下午茶
British Afternoon Tea

茶风篇 Spread in the World

中国茶——身份的象征
Chinese Tea – Status Symbol

中国茶叶在17世纪进入英国。由于当时科技的制约，航海技术落后，中国茶叶在英国的价格有时候甚至超过了黄金。在当时的英国，只有上层贵族才能享用到中国来的茶叶，所以中国茶成了一种身份与地位的象征。当时的英国上流社会中有个奇怪的现象，每个贵族妇女腰间都有一把精致的小钥匙，这并不是用来开房门或保险箱的，而是用来开启放茶叶的箱子，可见中国茶在当时的英国受到怎样的喜爱，同时又是何等的珍贵。

Chinese tea was introduced to Britain in the 17th century. Because of the restriction on a scientific and nautical technological level, the price of Chinese tea in Great Britain was more than gold. In Britain, only aristocratic families could enjoy Chinese tea. So Chinese tea became a status symbol at that time. A strange phenomenon began to occur in the upper class. Every noble lady had an exquisite minor key at herwaist. It was not for opening the door or the safe but for the tea storing case. So Chinese tea was trendy and precious.

离不开茶的英国人
Inseparable from Tea

英国人对茶的喜爱世界闻名，无论是悠闲的贵族，还是繁忙的上班族，总会有一些自己的时间停下来喝杯茶。不管是早餐茶、工休茶、中午茶还是下午茶，饮茶已经成为英国人的传统和特色。

Britain has become one of the best-known countries for its love of tea. No matter whether those leisurely nobles or busy office workers, they would stop to enjoy tea. Drinking tea has become a tradition and characteristic of Great Britain.

传统的英式下午茶总会配有放满精美的点心的三层篮。最下面一层一般是佐有火腿、熏鱼、黄瓜、青菜的三明治，中间是配有奶油或果酱的英式松饼，最上层是时令水果，通常从最下层开始取用。17世纪，安娜玛利亚公爵夫人带动了英国上层社会下午茶的流行。18世纪中期以后，下午茶开始走进普通人的生活，至今喝下午茶依然还是绝大多数英国人的习惯。

The traditional English ofternoon tea always a three-layer basket for setting some good snacks. Ham, smoked fish, and cucumber sandwiches would be put on the bottom layer. The middle layer has a toasted English muffin. The top layer has some fresh fruits in season. We usually enjoy the snacks from the bottom. The Duchess of Anna Maria made tea culture famous in the British upper class in the 17th century. Ordinary people started to enjoy afternoon tea after the middle of the 18 century. Afternoon tea has become a habit to this day.

一天从早餐茶开始
Awake from the Morning Tea

嗜茶的英国人每一天都是从茶开始的，这就是必不可少的"早餐茶"。在很多英国家庭中，一觉醒来后，家中的妇女就要在第一时间烧上水，准备为家人冲泡一杯早茶，这也是唤醒家人的最好方法。早餐茶也称"床茶"，适合清早醒来后空腹饮用，这种饮茶习惯被视为一种舒适的享受。据说，时至今日，英国女王一天的生活还是从喝早餐茶开始的。早餐茶通常精选印度、锡兰或肯尼亚各地的红茶调制而成，而正宗的中国茶多在下午茶时饮用。

British start the day with morning tea. The morning tea is deemed essential. Most British women will brew morning tea to wake families up. So morning tea is also called bed tea. Drinking tea on an empty stomach was considered a habit at that time. It is said that the queen of England still enjoys tea every day. They always select and blend

Black Tea from India, Sri Lanka, Kenya, and worldwide for morning tea. Traditional Chinese tea will be drunk in the afternoon.

下午茶的诞生
The Birth of Afternoon Tea

1840年,即维多利亚时代,有个叫做安娜玛利亚的公爵夫人,每到下午她都会感觉肚子饿,因为离吃晚餐还有一段时间,所以她喜欢让女仆准备好茶和点心,送进房间里美美地享受一番。由于公爵夫人每天对茶和点心的要求不断增加,女仆准备的茶点也越来越精致。每当有客人到访,公爵夫人总会盛情邀请对方一起品尝这晚餐前的美味,共同享受午后美好的时光。久而久之,下午茶成为英国贵族一种新的社交文化。随着时间的推移,就形成了当今英国社会流行的"英式午茶"。

In 1840, the Vicforian, there was a duches named Anna Maria, The Duchess of Anna Maria made tea culture famous in the British upper class in the 17th century. She felf hungry every afternoon at that time. Aa the Duchess's denands for tea and refreshments increased, the refreshmen prepared by the maid became more andwore exquisite. But the dinner was still a long time off. So she will ask her maids to prepare good tea and dessert and invite her guests to enjoy tea together. They wanted a good time in the afternoon. Over time, afternoon tea became a new social culture. As time goes by, British afternoon tea has been passed down.

19世纪，饮下午茶的风气在英国蔚然兴起。无论是贵族还是平民，无论在家中还是公共场所，由于茶点的加入，一套茶具中加入蛋糕盘和边盘才算完整。除此之外，还有蛋糕刀、茶点叉、松饼盘、餐巾、桌布、保暖罩、茶刀、滤器等辅助茶具。精美的细瓷茶具最为英国人所喜爱。另外，上流社会和贵族阶层也有使用银质茶具的。

There was a fashion of afternoon tea in Britain in the 19th century. A tea set is complete with the tea snake's plate. Some extra utensils include a cake knife, tea fork, muffin plate, napkin, tablecloth, heat shield, tea knife, and filter. The British love those fine porcelain tea sets. The upper class and noble class used silver tea sets.

摩洛哥茶饮
Morocco Tea Drinking

身体的一半是绿茶
Prefer Green Tea

摩洛哥地处非洲西北部地区，终年气候干燥炎热，又由于摩洛哥属穆斯林国家，当地人都不饮酒，所以便选择了茶这种既能提神，又可消暑、解渴的饮料作为日常饮品。茶对于摩洛哥人的重要性仅次于吃饭。摩洛哥是世界上进口绿茶最多的国家。在摩洛哥，大部分家庭的收入都用于饮茶，可见茶在摩洛哥人生活中的地位。所以摩洛哥人常常是说自己"身体的一半是绿茶"。

Morocco locates in Northwest Africa. It is dry and hot all year round. Besides, Morocco is a Muslim country, and locals don't drink wine. So they select tea as their ordinary beverage for refreshing themselves and reliving summer heat. Its importance is just inferior to the meal for Moroccans. Morocco is the biggest importer country of Green Tea in the world. Morocco is the biggest importer country of Green Tea in the world. Tea plays a vital role in their life. So most Moroccans always say that half of the body is Green Tea.

摩洛哥人的泡茶方法
Moroccans Brewing Method

摩洛哥的茶壶很奇特，一般的茶壶外表由铜浇铸而成，壶内镀上一层银，壶嘴很长，类似老北京功夫茶馆中的茶壶。茶杯雕刻着富有民族特色的图案、花纹，非常精致。

The Moroccan tea bowl is peculiar, made of copper with inner silver. It has a long spout opening and is similar to a teapot in the Gong Fu teahouse of old Beijing. Teacups are carved with elegant and beautiful patterns. They have characteristics and aesthetic perspectives.

摩洛哥人泡茶，首先将少许茶叶放于壶中，用沸水洗过一遍后倒掉，再冲上开水，并加上白糖。由于摩洛哥气候炎热，所以他们总会在茶壶中放入一大片薄荷叶，还可起到消暑的作用。最后盖上茶壶盖，静待几分钟后便可饮用。冲泡第二、第三次时，还要补充茶叶和糖。一壶茶通常可以冲泡三次，嗜甜的摩洛哥人需要茶叶与白糖的比例通常为1:10。

Moroccans put a few teas in the kettle. Then throw up the tea soup after being washed with boiled water. Pour boiled water again and add some white sugar. Because it is hot all year round, Moroccans always put some mint leaves in the tea soup. It can help to relieve summer heat. Cover the lid at last and wait for a few minutes. Then you can enjoy tea soup. You should add tea leaves and sugar for the second and third brewing. The tea can usually be brewed three times. Monacans have a sweet tooth. So the ratio of tea to sugar is 1:10.

经历了曲折的历程，随着一带一路的发展，中国茶这一曾经深

刻影响并引领世界的饮品，必将在现代继续勾画全球的饮品市场。我们以茶文化为媒，融合世界各地的文化，以文化做导引推广中国茶叶，为世界送上富有中国特色的文明体系。

After a tortuous process, with the development of the Belt and Road, Chinese tea, a beverage that once deeply influenced and led the world, will continue to glitter in the global beverage market. We use tea culture as a medium, integrate local cultures from around the world, and use culture as a guide to promote Chinese tea, providing the world with a civilization system rich in Chinese characteristics.

在一定的条件下，茶叶的采制工具和饮茶工具有时可以省略。

Under certain conditions, the use of tea-harvesting tools and tea-drinking tools can be omitted Somefimes.

若在初春禁火的时候，在野外寺院的山间茶园里，大家齐心齐力采摘，就地蒸捣，用火烘干，其他几种工具就可以不使用了。

In the early spring, people in the outdoor temple's mountain tea garden can work together to pick, steam, and pound on-site, dry with fire, and other tools can not be used at that time.

用四幅或六幅洁白的绢纸把《茶经》所述各项分别写在上面，陈列在座位旁。这样，对茶的起源、采茶、制作和煮茶所需要的器具，以及煮茶、饮茶的方法，茶的历史轶事、产地分布，还有在不同情况下制茶和煮茶所需器具的不同等，就可了解清楚。

Then, the origin of tea, tea picking, tea processing, tea brewing utensils, tea brewing, and drinking methods, historical anecdotes about tea, distribution of tea producing areas, and different situations for tea making and brewing in different situation and so on which can be clearly understood in the picfures.

自古以来，中国茶叶随着丝绸之路传到欧洲、逐渐风靡世界，它被认为是共结和平、友谊与合作的纽带。

Since ancient times, Chinese tea has spread to Europe along the Silk Road and gradually became famous in the world. It is considered as a peace, friendship, and cooperation bond.